Translations

of

Mathematical Monographs

Volume 37

Foundations of a Structural Theory
of Set Addition ₁ 1973.

by

G. A. Freiman

American Mathematical Society
Providence, Rhode Island
1973

НАЧАЛА СТРУКТУРНОЙ ТЕОРИИ СЛОЖЕНИЯ МНОЖЕСТВ

Г. А. ФРЕЙМАН

Под редакцией
А. М. Люстига и Л. П. Усольцева

КАЗАНСКИЙ ГОСУДАРСТВЕННЫЙ ПЕДАГОГИЧЕСКИЙ
ИНСТИТУТ

ЕЛАБУЖСКИЙ ГОСУДАРСТВЕННЫЙ ПЕДАГОГИЧЕСКИЙ
ИНСТИТУТ

КАЗАНЬ 1966

AMS (MOS) subject classifications 1970.
Primary 10–02, 10J99.

Library of Congress Cataloging in Publication Data

Freiman, G   A
    Foundations of a structural theory of set
addition.

    (Translations of mathematical monographs, v. 37)
    Translation of Nachala strukturnoĭ teorii
slozheniia mnozhestv.
    1. Numbers, Theory of. 2. Set theory.
I. Title. II. Series.
QA241.F7313   512'.7   73-9804
ISBN 0-8218-1587-3

Copyright © 1973 American Mathematical Society

TABLE OF CONTENTS

iii

PREFACE

The present book is addressed to students of graduate courses and to doctoral candidates studying number theory as well as to the specialist working in this field. As an experiment, the material was treated by a seminar on "Supplementary Topics in Number Theory" held at Elabuga State Pedagogical Institute, and it gives a systematic exposition of a series of papers on additive number theory by the author. Only a basic knowledge of number theory is presupposed, such as contained, for example, in the books by I.M. Vinogradov or A.A. Buhštab. Many of the problems treated in the first chapter may be used as topics for term papers or masters' theses as well as for independent mathematical research.

The last decade has seen the beginning of the application in additive number theory of general principles arising from the addition of sets. The present book follows this trend.

From 1959 to 1964 a series of papers (cf. [15] to [23]) were published under the general title, "Inverse problems of additive number theory". In these papers the structure of sets A with small "double set" $A + A$ was investigated. More precisely, if we consider a set with a certain numerical parameter (the number of elements for finite sets of numbers or of residue classes, the density for sequences of integers, the measure for point sets) for which that numerical parameter of the corresponding double set is not too large in comparison with the original value, then the structure of the original set can be described in a certain sense. The content of the papers mentioned is, by and large, the subject of this book.

The book consists of three chapters. The first chapter contains an exposition of the basic concepts, general ideas and elementary results. In the second chapter a difficult specific inverse problem is solved. The third chapter is devoted to applications.

I wish to stress the importance of the concept of isomorphism between subsets of sets with an algebraic operation as introduced and applied in Chapter I.

The well-known significance of the concept of isomorphism between groups lies in the fact that it provides the possibility for disregarding incidental properties of each given specific group, and thus it leads to investigations of a more general kind. Furthermore it permits us to introduce an infinity of operations on the elements of a group.

In additive number theory, on the other hand, one often considers sums consisting of a bounded number of terms from suitable sets. Therefore we restrict the investigations in the present book to problems concerned with simple (i.e. noniterated) addition. It is natural to introduce a concept of isomorphism which is geared to this situation.

There is an analogy between a local isomorphism of topological groups and an isomorphism of subsets, a so-called "algebraically local isomorphism". In the first case the isomorphism is restricted by the boundary of a certain neighborhood, and in the second case, by the number of operations performed.

The concept of isomorphism between subsets has permitted us to express a general view on additive number theory as the theory which studies those properties of sets of numbers which are preserved by such isomorphisms.

The author has endeavored to make the presentation of the material in the first chapter as explicit and elaborate as possible in order to introduce the beginning student to a new field of ideas.

The exercises are, by and large, concentrated in the first chapter. Some of them are strictly for teaching purposes whereas others provide material for independent research.

The material of the second and third chapters is presented more briefly than that of Chapter I.

In Chapter II a fundamental theorem on the structure of finite sets of integers with small "double set" is proved. In this proof we use a modification of the method of trigonometric sums and also the language and the elementary methods of the geometry of numbers.

It would be highly desirable to obtain generalizations of the results of Chapters I and II for arbitrary abelian and especially for non-abelian groups. The choice of the title for the present work was prompted by the desire to stress such possibilities of developing the theory along with the prospects for its further number-theoretical extensions.

Chapter III is devoted to applications of the theory.

As a summary we give a revised form of the English text read in September 1965 at the Summer School on Number Theory held in Palanga, Latvian SSR. In that paper a general exposition was given on investigations in the field of addition of finite sets of numbers which are of special significance in additive number theory.

EXPLANATIONS FOR THE USE OF THE BOOK

1. Definitions are given in the text whenever a new concept occurs for the first time. They may be located by means of the Index.

2. Whenever a concept known from the literature occurs in this book for the first time, it is printed in italics (sometimes the italics are repeated further on). References to the appropriate literature are given in the Index.

3. Subsections are numbered consecutively in each chapter, without regard to

sections. They are designated by two numbers, the first of which is the chapter number. The sign § is used for both sections and subsections; since sections (rarely referred to) are designated by only one number, this should cause no confusion. Thus, e.g., §1.10 is the tenth subsection of Chapter I; it happens to be part of §2 of that chapter.

4. References to theorems, lemmas and definitions are listed by the numbers of the chapter and the subsection in which they are given. Thus, e.g., Theorem 1.9 is to be found in §1.9 of Chapter I. Formulas are denoted by three numbers, i.e. the number of the chapter, the number of the subsection and the number of the formula. Thus, e.g., formula (2.3.1) is the first formula in §2.3 of Chapter II. The first number is omitted if reference is made to a formula of the same chapter. The first two numbers are omitted if reference is made to a formula of the same subsection. Thus, the reference to formula (1) on page 49 refers to formula (2.3.1) of the same subsection.

5. A number of problems of the caliber of exercises are given without asterisks, whereas the remaining ones, which are of a more advanced nature, are marked with one, two or three asterisks, depending on their difficulty.

CHAPTER I

ISOMORPHISMS

§1. Isomorphisms between subsets of sets with an algebraic operation

1.1. *Addition of sets*. Let E be a set with an *algebraic operation*, written as addition, and let B and C be subsets of E.

The sum $B + C$ is defined as the subset of E consisting of those elements x which possess at least one representation in the form $x = b + c$, $b \in B$, $c \in C$.[1]

It is obvious how this definition extends to the case of the addition of an arbitrary number of subsets.

The sum $B + B$ of two identical subsets shall be denoted by $2B$.

EXAMPLES. 1. Let N_2 be the set of nonnegative even numbers, $\{0, 1\}$ the set consisting of the numbers 0 and 1, and let N be the set of positive integers. Then $N_2 + \{0, 1\} = N \cup \{0\}$; $2N_2 = N_2$.

2. Let Q_2 be the set of squares of the integers. Then Lagrange's theorem on the representation of each positive integer as the sum of at most four squares is expressible in the form $4Q_2 = N \cup \{0\}$.

3. Let A be an abelian group. Then we have $2A = A$ and, in particular, $2Z = Z$, where Z is the additive group of integers. Let B and C be two sets with an algebraic operation defined on each of them in such a way that these two structures are isomorphic. The significance of the concept of isomorphism in the investigation of an algebraic operation consists in the fact that we may disregard specific properties of the set on which the algebraic operation is defined by substituting for it any other set which is isomorphic to the given one.

In defining an isomorphism one takes into account the possibility of introducing any finite number of operations on the elements of a set. Thus, if $b_i \in B$, $c_i \in C$, and b_i corresponds to c_i under the isomorphism, written as $b_i \leftrightarrow c_i$, $1 \leqslant i \leqslant n$, where n is any positive integer, then $\sum_{i=1}^n b_i \leftrightarrow \sum_{i=1}^n c_i$. If mutually disjoint subsets B' and C' of the sets B and C are given, then the latter relation may be expressed in the form

$$nB' \leftrightarrow nC', \qquad n = 1, 2, \ldots. \tag{1.1.1}$$

---

[1] The algebraic sum $B + C$ thus defined is not to be confused with the set-theoretical union (sometimes called sum also) of the subsets B and C, which shall be denoted by $B \cup C$.

However, the relation (1) turns out to be too restrictive for sums of a finite number of identical sets as they occur frequently in additive number theory. In the present book we limit the exposition to problems related to sums of only two identical sets.

If we consider the remarks made above, we are led to the idea that in the case where the subsets B' and C' are each added once to themselves, they should be considered as isomorphic if condition (1) is satisfied for $n = 1$ and 2, but not necessarily for all values of n. A rigorous definition is given in the following subsection.

1.2. *Isomorphisms of subsets.*

DEFINITION. The subsets B' and C' of the sets B and C with an algebraic operation (written as addition) are called *isomorphic* if there exists a one-to-one mapping from B' onto C' $(B' \to C')$ which induces naturally a one-to-one mapping from $2B'$ onto $2C'$. The mapping $B' \to C'$ is then called an *isomorphism*.

We give examples and explanations concerning this definition.

How is the mapping $2B' \to 2C'$ defined? We say that the mapping $2B' \to 2C'$ is naturally induced by the mapping $B' \to C'$ if, for any $b_1, b_2 \in B'$ and $c_1, c_2 \in C'$ with $b_1 \to c_1$ and $b_2 \to c_2$, the element $b_1 + b_2$ from $2B'$ is mapped onto the element $c_1 + c_2$ from $2C'$.

EXAMPLES OF ISOMORPHIC SETS. 1. Let B' and C' be subsets of the sets B and C with an algebraic operation. If B' and C' consist of two elements each, then they are isomorphic.

2. The subsets $\{0, 1, 3\}$ and $\{0, 1, 5\}$ of the additive group of integers are isomorphic.

3. The subset $\{0, 1, 3\}$ of the additive group of integers and the subset $\{(0, 0), (1, 0), (0, 1)\}$ of the additive group of integral sectors in the plane are isomorphic to each other.

4. Let S_p be the additive group of residue classes modulo p. The set of integers $\{0, 1, 2, 3\}$ is isomorphic to the set $\{0, 1, 2, 3\}$ of residues of the group S_7 but not isomorphic to the set $\{0, 1, 2, 3\}$ of residues from S_5.

5. The semigroups $B' = \{0, 1, 2, \ldots\}$ and $C' = \{10, 11, 12, \ldots\}$ within the additive group of integers are isomorphic.

The possibility of performing a one-to-one mapping $B' \to C'$ does not necessarily imply that B' and C' should be isomorphic, since no induced mapping $2B' \to 2C'$ needs to exist.

EXAMPLE. Let $B' = \{b_1, b_2, b_3\}$ and $C' = \{c_1, c_2, c_3\}$, where B' and C' are sets of integers, numbered in the order of increasing magnitude. Let $b_1 + b_3 = 2b_2$ and $c_1 + c_3 \neq 2c_2$. Furthermore, let $B' \to C'$ be an arbitrary one-to-one mapping of B' onto C'. For example, we may have $b_i \to c_i$, $1 \leqslant i \leqslant 3$ (the remaining cases are treated analogously). Then it follows that $2b_2 \to 2c_2$ and $b_1 + b_3 \to c_1 + c_3$.

Thus one and the same element $b_1 + b_3 = 2b_2$ of the set $2B'$ must be mapped onto either of the two different elements $c_1 + c_3$ and $2c_2$ of the set $2C'$, which contradicts the definition of the mapping.

It may happen that the mapping $2B' \to 2C'$ exists but fails to be one-to-one.

EXAMPLE. We consider the mapping $C' \to B'$ with the sets defined as in the preceding example. In this case the mapping $2C' \to 2B'$ exists. However, for example, the mapping $c_i \to b_i$ maps both $c_1 + c_3$ and also $2c_2$ onto the same element, and thus the mapping $2C' \to 2B'$ cannot be one-to-one.

In the last two examples the number of elements of the set $2B'$ was not equal to the number of elements of the set $2C'$ (see, however, Exercise 1, below).

The definition of an isomorphism between subsets may also be worded in the following way.

Let B and C be sets with an algebraic operation defined on both of them, and let B' and C' be given subsets of B and C, respectively. Then B' and C' are called *isomorphic* if the following statements are true:

1) There exists a one-to-one correspondence between B' and C'.

2) There exists a one-to-one correspondence between $2B'$ and $2C'$ such that the sum of any two elements from B' is mapped onto the sum of their images in C' and vice versa; that is if $b_1, b_2 \in B'$, $c_1, c_2 \in C'$, $b_1 \leftrightarrow c_1$ and $b_2 \leftrightarrow c_2$, then $b_1 + b_2 \leftrightarrow c_1 + c_2$. Suppose the mappings $B' \to C'$ and $2B' \to 2C'$ correspond to each other, i.e. any element of the subsets B' and $2B'$ is mapped onto the same element by either of these transformations.

If furthermore $B' = B$ and $C' = C$, then the definition given above reduces to the general definition of an isomorphism of sets with an algebraic operation, since in this case $2B' \subset B'$ and $2C' \subset C'$.

EXERCISES. 1) Exhibit nonisomorphic sets B' and C' each of which consists of 4 integers such that the sets $2B'$ and $2C'$ both have the same number of elements.

2) Show that in the group S_5 any two subsets with three elements are isomorphic.

3) Let A be a subset of the group S_5 consisting of k elements and show that for $k \leqslant 3$ there exists an isomorphism of the set A into the additive group Z of integers. For $k > 3$ no such mapping exists.

4) Show that there does not exist an automorphism of the group Z which maps the sets of Example 2 onto each other. For which n is the relation (1.1) violated?

5) Verify that the mappings $B' \to C'$ and $2B' \to 2C'$ in Example 5 do not correspond to each other.

1.3. *Isomorphisms of order s.* The definition of an isomorphism between subsets as given in §1.2 refers to single (i.e. not iterated) addition of sets. Thus it forms a special case (for $s = 2$) of the definition given below.

DEFINITION. The subsets B' and C' of B and C with an algebraic operation are called *isomorphic of order s* if the mapping $sB' \to sC'$ which is naturally induced by a one-to-one mapping $B' \to C'$ exists and is one-to-one. The mapping $B' \to C'$ is then called an isomorphism of order s. Sets which are isomorphic in the sense of §1.2 may consequently be called isomorphic of order two. We shall not use this term inasmuch as we shall only consider single addition of sets in the sequel. It is possible for an isomorphism of order s to exist with none of order $s + 1$ (see Exercise 1).

It is obvious from (1.1) that for any positive integer s an isomorphic mapping of a set B with an algebraic operation onto a set C induces an isomorphic mapping of order s of any subset $B' \subset B$ onto the appropriate image set $C' \subseteq C$.

EXERCISES. 1) Show that the sets of integers $K_1 = \{0, 1, s + 1\}$ and $K_2 = \{0, 1, s + 2\}$, $s \geqslant 1$, are isomorphic of order s but not of order $s + 1$.

2*) Let B' and C' be subsets of the sets B and C with an algebraic operation defined on each of them. Does an isomorphism of order s imply an isomorphism of order $s - 1$?

1.4. *Isomorphism between two finite classes of sets.* Finally we give a definition of isomorphism which is adapted to the case of addition of sets which are not necessarily equal to each other.

DEFINITION. Suppose two sets B and C are given with an algebraic operation defined on either of them, and consider two finite classes of subsets $B_i \subset B$ and $C_i \subset C$, $1 \leqslant i \leqslant s$. The class $\{B_i\}$ is called *isomorphic to the class* $\{C_i\}$ if there exists a one-to-one mapping $B_i \to C_i$, $1 \leqslant i \leqslant s$, such that there exists a naturally induced one-to-one mapping $B_1 + B_2 + \ldots + B_s \to C_1 + C_2 + \ldots + C_s$.

The definitions of §§1.2 and 1.3 are special cases of this general definition.

1.5. *A criterion for isomorphism.* If the sets B and C are abelian groups then the following simple criterion for isomorphism between subgroups holds.

THEOREM. *Let B' and C' be subsets of the abelian groups B and C, respectively. Furthermore, let a one-to-one correspondence between B' and C' be given in such a way that $b_1 - b_2 = b_3 - b_4$ implies $c_1 - c_2 = c_3 - c_4$ and that $b_1 - b_2 \neq b_3 - b_4$ implies $c_1 - c_2 \neq c_3 - c_4$ whenever $b_i \in B'$, $c_i \in C'$ and $b_i \leftrightarrow c_i$, $1 \leqslant i \leqslant 4$. This is a necessary and sufficient condition in order that the subsets B' and C' be isomorphic.*

We show that the conditions of the theorem imply that B' and C' are isomorphic. In order to establish this it suffices to show that the mapping $2B' \to 2C'$ naturally induced by the mapping $B' \to C'$ is one-to-one. This follows from the fact that if $b_1 + b_4 = b_2 + b_3$ then $b_1 - b_2 = b_3 - b_4$, and this implies $c_1 - c_2 = c_3 - c_4$ and thus $c_1 + c_4 = c_2 + c_3$. This conclusion remains true if equality is replaced by inequality in all four places.

Now suppose that B' is isomorphic to C'. If $b_1 + b_4 = b_2 + b_3$, i.e. some element from $2B'$ has the two representations $b_1 + b_4$ and $b_2 + b_3$, then the corresponding element from $2C'$ has the two representations $c_1 + c_4$ and $c_2 + c_3$, i.e. $c_1 + c_4$

$= c_2 + c_3$. But if $b_1 + b_4 \neq b_2 + b_3$ then also $c_1 + c_4 \neq c_2 + c_3$ and hence the equation $c_1 + c_4 = c_2 + c_3$ implies $b_1 + b_4 = b_2 + b_3$.

EXERCISES. 1) Exhibit an isomorphism between the sets of Examples 2, 3 and 4 of §1.2 by means of the criterion for isomorphisms.

2) Any arithmetic progression consisting of m real numbers, considered as a subset of the additive group of real numbers, is isomorphic to the set $\{0, 1, 2, \ldots, m - 1\}$.

3) Prove Theorem 1.6 for the case $k = 3$.

4) A linear transformation of the additive group D of real numbers induces an isomorphic transformation of any finite subset of the group D.

1.6. *The number of classes of isomorphic sets.* By K we shall denote a finite subset of an abelian group consisting of k elements. In this subsection we assume $K \subset D$, whereas in §2.1, 2.3, 3.12 and 3.13 we shall have $K \subset S_p$, and in the remaining cases we shall assume $K \subset Z_n$, where Z_n is the additive group of integral vectors of the euclidean space E_n.

Thus, $K = \{\bar{a}_0, \bar{a}_1, \ldots, \bar{a}_{k-1}\}$, where $\bar{a}_i$ are integral vectors in the case where $K \subset Z_n$. In the case where $K \subset D$ we have $K = \{a_0, a_1, \ldots, a_{k-1}\}$ with real numbers a_i for which we may always assume that $a_i < a_{i+1}$, $i = 0, 1, \ldots, k - 2$.

We consider the additive group D of real numbers and in it a finite subset K consisting of k numbers. Isomorphism between two such subsets is an *equivalence relation* (it is reflexive, symmetric and transitive). Thus, the class of all sets of real numbers may be decomposed into *classes of mutually isomorphic sets*. The following theorem describes these classes for $k = 3$, 4 and 5.

THEOREM. *Consider a finite subset K, consisting of k elements, of the additive group of real numbers. An arbitrary set consisting of three real numbers is isomorphic to one of the two sets $K_{1,3} = \{0, 1, 2\}$ or $K_{2,3} = \{0, 1, 3\}$. An arbitrary set consisting of four elements is isomorphic to one of the following five sets: $K_{1,4} = \{0, 1, 2, 3\}$, $K_{2,4} = \{0, 1, 2, 4\}$, $K_{3,4} = \{0, 1, 2, 5\}$, $K_{4,4} = \{0, 1, 3, 4\}$, $K_{5,4} = \{0, 1, 3, 7\}$. An arbitrary set consisting of five elements is isomorphic to one of the following twenty-two sets: $K_{1,5} = \{0, 1, 2, 3, 4\}$, $K_{2,5} + \{0, 1, 2, 4, 6\}$, $K_{3,5} = \{0, 2, 3, 4, 6\}$, $K_{4,5} = \{0, 1, 2, 3, 5\}$, $K_{5,5} = \{0, 1, 2, 3, 6\}$, $K_{6,5} = \{0, 1, 2, 3, 7\}$, $K_{7,5} = \{0, 1, 2, 4, 5\}$, $K_{8,5} = \{0, 1, 2, 4, 7\}$, $K_{9,5} = \{0, 1, 2, 4, 8\}$, $K_{10,5} = \{0, 3, 4, 5, 7\}$, $K_{11,5} = \{0, 4, 5, 6, 8\}$, $K_{12,5} = \{0, 2, 4, 5, 8\}$, $K_{13,5} = \{0, 1, 2, 4, 9\}$, $K_{14,5} = \{0, 3, 4, 5, 10\}$, $K_{15,5} = \{0, 3, 4, 5, 8\}$, $K_{16,5} = \{0, 1, 2, 4, 6\}$, $K_{17,5} = \{0, 1, 2, 5, 7\}$, $K_{18,5} = \{0, 4, 5, 6, 9\}$, $K_{19,5} = \{0, 1, 2, 5, 8\}$, $K_{20,5} = \{0, 1, 2, 5, 11\}$, $K_{21,5} = \{0, 1, 3, 4, 0\}$, $K_{22,5} = \{0, 1, 3, 7, 15\}$.*

PROOF. If $k = 3$, then either K is an arithmetic progression and thus isomorphic to $K_{1,3}$, or it is not an arithmetic progression and thus isomorphic to $K_{2,3}$.

If $k = 4$ and K is an arithmetic progression then $K \sim K_{1,4}$ ($\sim$ denotes isomorphism). Now suppose there exists in K an arithmetic progression consisting of three terms, i.e. $\{a, a + q, a + 2q\} \subset K$. If one of the numbers $a + 4q$, $a - 2q$, $a + q/2$, $a + 3q/2 \in K$, then $K \sim K_{2,4}$; otherwise we have $K \sim K_{3,4}$. If there does not exist in K any arithmetic progression of three terms but there exist identical differences then $K \sim K_{4,4}$. If no identical differences exist, then $K \sim K_{5,4}$.

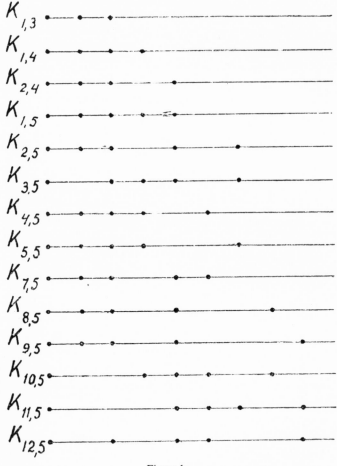

Figure 1a

Now let $k = 5$. If K is an arithmetical progression then $K \sim K_{1,5}$. Now let there exist in K an arithmetical progression consisting of four numbers, i.e. $\{a, a + q, a + 2q, a + 3q\} \subset K$. If $a + q/2$ or $a + 5q/2$ are contained in K then $K \sim K_{2,5}$; if $a + 3q/2 \in K$ then $K \sim K_{3,5}$; if $a + 5q$ or $a - 2q$ belong to K then $K \sim K_{4,5}$; if

$a + 6q$ or $a - 3q$ are contained in K then $K \sim K_{5,5}$. In the remaining cases we have $K \sim K_{6,5}$.

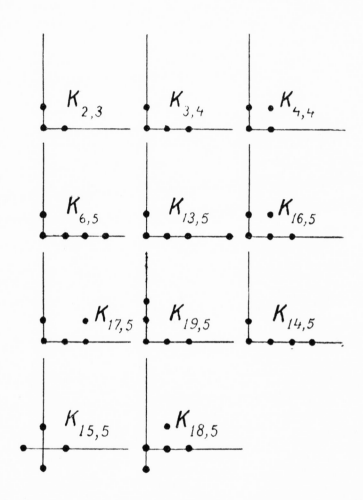

Figure 1b

We consider the case where there exists in K a subset isomorphic to $K_{2,4}$ but no arithmetic progression consisting of four terms. Let $\{a, a + q, a + 2q, a + 4q\} \subset K$. If $a + 5q \in K$ then $K \sim K_{7,5}$; if $a + 7q \in K$ then $K \sim K_{8,5}$; if $a + 8q \in K$ then $K \sim K_{9,5}$; if $a - 3q \in K$ then $K \sim K_{10,5}$; if $a - 4q \in K$ then $K \sim K_{11,5}$; if $a + q/2 \in K$ then $K \sim K_{9,5}$; if $a + 3q/2 \in K$ then $K \sim K_{11,5}$; if $a + 5q/2 \in K$ then $K \sim K_{12,5}$. In all other cases one has $K \sim K_{13,5}$.

Now suppose that there exists in K a subset $K_3 = \{a, a + q, a + 2q\}$ but no subset which is isomorphic to either $K_{1,4}$ or $K_{2,4}$. Suppose the set K also contains the numbers b and c. We assume that among the numbers of the sets

$$K_3 + \{-b\}, \; K_3 + \{-c\} \tag{1.6.1}$$

some of them coincide in absolute value.

If $a - b = c - a$ then $K \sim K_{14,5}$; if $a + q - b = c - a - q$ then $K \sim K_{15,5}$; if $a - b = a + q - c$ then $K \sim K_{16,5}$; if $a - b = a + 2q - c$ then $K \sim K_{17,5}$; if $a - b = c - a - q$ then $K \sim K_{18,5}$. The remaining cases with equality of absolute values do not lead to new sets. Now suppose that among the numbers of the sets (1) there does not exist any equality of absolute values, but the number $b - c$ is equal in absolute value to one of the numbers of the sets (1). If $a - b = b - c$ then $K \sim K_{19,5}$. The remaining cases do not lead to new sets. Finally, if $b - c$ is not equal in absolute value to any of the numbers (1) then $K \sim K_{20,5}$.

Now suppose that K does not contain any set isomorphic to $K_{1,3}$. If there exist equal differences in K then $K \sim K_{21,5}$, and if no such equal differences exist then $K \sim K_{22,5}$.

Figures la and lb show isomorphic images of the sets of Theorem 1.6. We denote by $t(k)$ the number of classes of isomorphic sets consisting of k real numbers. It follows from Theorem 1.6 that $t(1) = 1$, $t(2) = 1$, $t(3) = 2$, $t(4) = 5$ and $t(5) = 22$.

In Table 1 we list the values T for the sets of Theorem 1.6 and also some additional numerical parameters referring to §§1.18 and 1.28.

EXERCISES. 1*. Describe two classes of isomorphic sets of real numbers and determine $t(k)$ for $k = 7$.

2*. Establish specific numerical bounds for $t(k)$ for small values of k ($7 \leqslant k \leqslant 10$).

3. Prove that the function $t(k)$ is finite.

4**. Determine the order of growth of the function $t(k)$.

5*. Consider a finite subset consisting of k elements contained in an arbitrary abelian group. Generalize the results and the wording of the exercises from §1.6 and Exercises 1–4 to this case.

6*. The same for nonabelian groups.

§2. Addition of finite sets. Elementary results

1.7. *Direct and inverse problems of additive number theory.* Classical additive number theory deals with the representation of positive integers as sums of terms of a specified form. Usually certain conditions are imposed on the number of terms. The fundamental problems are to determine which numbers can be so represented and to give bounds on the number of representations.

K	k	T	r	M'	R
$K_{1,3}$	3	5	1	5	2
$K_{2,3}$	"	6	2	3	3
$K_{1,4}$	4	7	1	14	3
$K_{2,4}$	"	8	1	10	4
$K_{3,4}$	"	9	2	8	5
$K_{4,4}$	"	9	2	10	4
$K_{5,4}$	"	10	3	6	6
$K_{1,5}$	5	9	1	30	4
$K_{2,5}$	"	11	1	20	6
$K_{3,5}$	"	11	1	22	5
$K_{4,5}$	"	10	1	24	5
$K_{5,5}$	"	11	1	20	6
$K_{6,5}$	"	12	2	18	7
$K_{7,5}$	"	11	1	22	5
$K_{8,5}$	"	12	1	16	7
$K_{9,5}$	"	12	1	16	7
$K_{10,5}$	"	12	1	18	6
$K_{11,5}$	"	12	1	16	7
$K_{12,5}$	"	12	1	16	7
$K_{13,5}$	"	13	2	14	8
$K_{14,5}$	"	13	2	14	8
$K_{15,5}$	"	13	2	18	6
$K_{16,5}$	"	12	2	20	6
$K_{17,5}$	"	13	2	16	7
$K_{18,5}$	"	13	2	14	7
$K_{19,5}$	"	13	2	14	8
$K_{20,5}$	"	14	3	12	9
$K_{21,5}$	"	14	3	14	8
$K_{22,5}$	"	15	4	10	10

Table 1

A typical example of this question is Waring's problem, where the terms to be added are the powers (with fixed exponent) of the positive integers.

We shall call *direct additive problems* those problems in which the sets of terms are given and properties of the sum-sets are investigated.

By an *inverse problem* of additive number theory we mean a problem which calls for establishing certain facts about given sets of numbers by using information on their sum-sets.

One may consider a direct problem as analogous to the problem of finding the sum of two numbers if the summands are given, and an inverse problem as analogous to the problem of finding the two summands if their sum is given and certain additional stipulations are made.

If we do not insist on the arithmetical nature of the sets involved we may consider a structural theory of sum-sets which contains the inverse problem as a special part. The term "inverse problem of additive number theory" was introduced in 1955 in the papers [14] and [15]. We formulate an inverse problem for which a solution in principle will be given in Chapters 1 and 2 of the present book.

1.8. *Formulation of an inverse problem.* If M is a finite set, the number of its elements shall be denoted by $T(M)$.

Let $K = \{a_i\}$, $0 \leqslant i \leqslant k - 1$, where the a_i are integers, $a_i < a_{i+1}$, $i = 0, 1, \ldots, k - 2$, and consider the quantity $T = T(2K)$.

What can be said about the quantity T? One has the inequality

$$2k - 1 \leqslant T \leqslant k(k + 1)/2 \qquad (1.8.1.)$$

The validity of the lower bound for T follows from the fact that, for any such set K, the set $2K$ contains the following $2k - 1$ different numbers: $2a_0, a_0 + a_1$, $a_0 + a_2, \ldots, a_0 + a_{k-1}, a_1 + a_{k-1}, a_2 + a_{k-1}, \ldots, 2a_{k-1}$.

The validity of the upper bound for T follows from the fact that there are no more than $(k - 1)k/2$ sums of mutually different numbers from K along with no more than k sums of equal numbers.

The inverse problem consists in determining the structure of the set K if T is given.

We shall see that for small values of T the structure of K is always the same and can be well described.

In solving direct problems we are generally interested in the possibility of obtaining sharp lower bounds for T. If we can show that a given set K has a structure different from that which obtains for sets with small T, then we are at the same time establishing a lower bound for the number T belonging to this set K.

Thus, the solution of an inverse problem of the type formulated here provides a natural approach to the solution of a direct problem.

Exercises. 1. Show that the bounds given in (1) may occur.

2. A set K of integers with $T = k(k + 1)/2$ is isomorphic to the set of vertices of a tetrahedron in the space E_{k-1}.

1.9. *The structure of K for $T < 3k - 3$.* It turns out that for very small values of T (i.e. for $T < 3k - 3$,) the set K is contained in an arithmetical progression. Thus, $T = 2k - 1$ if and only if K is an arithmetical progression consisting of k terms. Indeed, if we have

$$a_{j+1} - a_j \neq a_{j+2} - a_{j+1}$$

for some j with $0 \leqslant j \leqslant k - 3$, then it is possible to add to the distinct numbers $2a_0, a_0 + a_1, 2a_1, a_1 + a_2, 2a_2, \ldots, 2a_{k-1}$ the number $a_j + a_{j+2}$ which is different from all of them.

The validity of the following theorem may be established by induction on k:

THEOREM. *If b satisfies $0 \leqslant b < k - 2$ and $T = 2k - 1 + b$, then the set K is a subset of a set K_a of the form*

$$K_a = \{a, a + q, a + 2q, \ldots, a + (k + b - 1)q\},$$

where a and q are integers with $a > 0$.

Thus, if $T < 3k - 3$, then the set K is contained in an arithmetic progression of no more than $T - k + 1$ terms.

Before we give the proof of this theorem (§1.10) we want to make some observations.

Since any two arithmetic progressions with the same number of terms are isomorphic (Exercise 2 of §1.5), Theorem 1.9 describes the structure of the set K up to an isomorphism.[2] In general this situation is typical for our inverse problem inasmuch as isomorphic sets K have the same value of T. From each class of isomorphic sets satisfying the conditions of the inverse problem we may therefore without loss of generality select some subclass defined by conditions given in advance.

Addition of a fixed number to all elements of a given set and also multiplication of all elements of a given set by a fixed nonzero number yields a set which is isomorphic to the given one (see Exercise 4 of §1.5). Therefore we may presuppose without loss of generality that $a_0 = 0$ and $d(K) = 1$, where $d(K)$ is the greatest common divisor

$$(a_1 - a_0, a_2 - a_0, \ldots, a_{k-1} - a_0)$$

of the numbers $a_1 - a_0, a_2 - a_0, \ldots, a_{k-1} - a_0$. In Theorem 1.9 we obtain under these conditions $a_0 = 0$, $q = 1$ and $a_{k-1} \leqslant k + b - 1$, and thus the assertion may be reworded as follows:

If $a_0 = 0$ and $d(K) = 1$ for the set K, then the condition $a_{k-1} \leqslant k + b - 1$ is satisfied provided that $0 \leqslant b < k - 2$ and $T = 2k - 1 + b$.

---

[2] This means that if a set K satisfies the conditions of the theorem then any set isomorphic to K also satisfies these conditions. We observe that, generally speaking, there exist nonisomorphic sets with the same values of T (see Exercise 1 of §1.2).

1.10. Proof of Theorem 1.9. We introduce still another equivalent version of Theorem 1.9.

Theorem. *Let $K = \{a_0, a_1, \ldots, a_{k-1}\}$, where the a_i are integers, $a_0 = 0$, $d(K) = 1$, and $a_i < a_{i+1}$, $i = 0, 1, 2, \ldots, k - 2$. If $a_{k-1} \geqslant k + b$, where b is an integer satisfying $0 \leqslant b < k - 2$, then $T \geqslant 2k + b$.*

Proof. First we show that for $a_{k-1} = k + b$ with $0 \leqslant b < k - 2$ the inequality $T \geqslant 2k + b$ holds. Let s be an arbitrary integer, $s \notin K$, $1 \leqslant s \leqslant k + b$. The number of such integers s is $b + 1$. Let a_j be the largest of the numbers a_i less than s. The numbers

$$a_{k-1} + s - a_i, \qquad 1 \leqslant i \leqslant k - 2 \qquad (1.10.1)$$

as well as the numbers

$$a_{j+1}, a_{j+2}, \ldots, a_{k-1}, a_{k-1} + a_1, a_{k-1} + a_2, \ldots, a_{k-1} + a_j \qquad (1.10.2)$$

are contained among the $a_{k-1} - 1$ consecutive numbers $s + 1, s + 2, \ldots, s + a_{k-1}$. In general the number of integers (1) and (2) is equal to $2k - 3$. Since $a_{k-1} - 1 \leqslant 2k - 4$, one of the number (1) must coincide with one of the numbers (2). This means that either $s \in 2K$ or $s + a_{k-1} \in 2K$. By taking into account the elements of the set K and $a_{k-1} + K$ we obtain

$$T \geqslant 2k - 1 + b + 1 = 2k + b. \qquad (1.10.3)$$

In order to complete the proof it remains now to show that for $a_{k-1} \geqslant 2k - 2$ the inequality $T \geqslant 3k - 3$ holds. For $k = 3$ this inequality is true (with $T = 6$). Suppose it has been established for all numbers from 3 up to $k - 1$ with $k \geqslant 4$.

First we consider the case $d(K') > 1$ where $K' = K \setminus a_{k-1}$. In this case the numbers $a_{k-1} + a_s$, $s = 0, 1, \ldots, k - 1$, do not belong to the set $2K'$. Therefore, letting $T' = T(2K')$, we obtain $T \geqslant T' + k \geqslant 2(k - 1) - 1 + k = 3k - 3$.

Thus we may assume in the sequel that $d(K') = 1$.

1) Let $a_{k-1} - a_{k-2} = 1$ and $a_{k-1} - a_{k-3} > 2$. In this case the numbers $a_{k-1} + a_s$, $s = k - 3, k - 2, k - 1$ do not belong to the set $2K'$. Thus by the induction hypothesis we have $T' \geqslant 3k - 6$, and therefore $T \geqslant T' + 3 \geqslant 3k - 3$.

2) Let $a_{k-1} - a_{k-2} \geqslant 2$ and $a_{k-2} \geqslant 2k - 4$. We choose j such that $a_{k-1} \equiv a_s \pmod{a_{k-1} - a_{k-2}}$ for $s = j + 1, j + 2, \ldots, k - 2$ and $a_{k-1} \not\equiv a_j \pmod{a_{k-1} - a_{k-2}}$. In this case the numbers $a_{k-1} + a_j$, $a_{k-1} + a_{k-2}$ and $2a_{k-1}$ do not belong to the set $2K'$.

We show that this is also the case for the number $a_{k-1} + a_j$. Otherwise we would have an equation of the form

$$a_{k-1} + a_j = a_s + a_t, \qquad j < s, t < k - 1,$$

from which we obtain

$$a_{k-1} - a_j = a_{k-1} - a_s + a_{k-1} - a_t.$$

But this is impossible since the right-hand side of the equation is divisible by $a_{k-1} - a_{k-2}$ whereas the left-hand side is not.

3) Let $a_{k-2} < 2k - 4$.

3a) $a_i < 2i$, $i = 1, 2, \ldots, k - 2$. Suppose $s \notin K$, $1 \leqslant s \leqslant 2k - 4$, and let a_j be the largest of the numbers a_i less than s. Considering the numbers a_i and $s - a_i$, $1 \leqslant i \leqslant j$, we find that $s \in 2K$ as in the proof of (3). Considering 0 and the k elements $a_{k-1} + a_s$, $s = 0, 1, \ldots, k - 1$, we obtain $T \geqslant 2k - 4 + 1 + k = 3k - 3$.

3b) Suppose there exists a number j with

$$a_{j-1} \geqslant 2j - 2, a_s < 2s, \qquad s = j, j + 1, \ldots, k - 2.$$

Since $a_{j-1} \geqslant 2j - 2$ and $a_j \leqslant 2j - 1$, we have

$$a_{j-1} = 2j - 2, \qquad a_j = 2j - 1.$$

Now we consider the two sets

$$\bar{K} = \{0, a_1, \ldots, a_{j-1}, a_j\} \tag{1.10.4}$$

and

$$\bar{\bar{K}} = \{a_{j-1}, a_j, \ldots, a_{k-1}\}. \tag{1.10.5}$$

An application of the inequality (3) to the set $\bar{K}$ for $k = j + 1$ and $b = j - 2$ yields $T(2\bar{K}) \geqslant 3j$. The induction hypothesis implies that $T(2\bar{\bar{K}}) \geqslant 3(k - j)$.

In view of the fact that the sums $2a_{j-1}$, $a_{j-1} + a_j$ and $2a_j$ are each counted twice, we obtain

$$T \geqslant T(2\bar{K}) + T(2\bar{\bar{K}}) - 3 \geqslant 3k - 3.$$

We observe that instead of the set K we may consider the set $K^* = \{0, a_{k-1} - a_{k-2}, a_{k-3}, \ldots, a_{k-1} - a_1, a_{k-1}\}$. Applying this remark to the remaining cases $a_{k-1} - a_{k-2} = 1$ and $a_{k-1} - a_{k-3} = 2$, we obtain a set for which $a_1 = 1$ and $a_2 = 2$. If we exclude sets K with such properties as we have already considered in the cases 1)–3), only the following case remains to be considered:

4) $a_2 = 2$ and $a_{k-1} - a_{k-3} = 2$. In this case there exists an integer j such that $a_{j-1} < 2j - 2$ and $a_j \geqslant 2j$. The proof is completed as in case 3b).

An investigation of the example

$$K = \{0, 1, \ldots, k - 2, k - 1 + b\}, \qquad 0 \leqslant b < k - 2, \qquad (1.10.6)$$

with $T = 2k - 1 + b$ shows that the theorem is best possible.

EXERCISES. 1. Exhibit all classes of isomorphic sets of integers with $T = 2k$, $k \geqslant 4$.

2. The same for $T = 2k + 1$, $k \geqslant 5$.

3*. Generalize the results of Theorem 1.9 to the plane (and more generally, to n-dimensional euclidean space).

4*. Determine the classes of sets of integers which are isomorphic of order s, with $T(sK) = sk - s + 1 + b$, where b is a small nonnegative integer and k is sufficiently large.

5*. Formulate and prove a generalization of Theorem 1.9 to sums of s identical sets of integers.

1.11. *More on the structure of the set K.* Are the essential features of the structure of K, as described by Theorem 1.9 for $T < 3k - 3$, preserved in the case when $T \geqslant 3k - 3$? More precisely, if we consider all arithmetic progressions which contain the set K and choose from them the arithmetic progression with the smallest number of terms, then that number of terms may be called "the length" of the set K. Is it then possible, considering the class of all sets K with given k and T, to give a bound for their length which depends on k and T but is independent of the set K? As the example

$$K = \{0, 1, 2, \ldots, k_1 - 1, b, b + 1, \ldots, b + k_2 - 1\},$$
$$k_1 + k_2 = k, \quad k_1, k_2 \geqslant 1, \quad b + k_2 \geqslant 2k, \qquad (1.11.1)$$

shows, this is not yet the case for $T = 3k - 3$, since b may be chosen arbitrarily large.

Thus it seems that the sets K with $T < 3k - 3$ and those with $T \geqslant 3k - 3$ are essentially different in structure, and everything which we shall learn about them in the sequel, and which shall be described in general terms, may be illustrated by considering the case $T = 3k - 3$.

The set K of the example (1) is isomorphic to the following set $K_0 \subset Z_2$:

$$K_0 = \{(0, 0), (0, 1), (0, 2), \ldots, (0, k_1 - 1), (1, 0), (1, 1), \ldots, (1, k_2 - 1)\},$$
$$k_1 + k_2 = k, \qquad k_1, k_2 \geqslant 1;$$

furthermore it has a length greater than $2k - 1$. It turns out that the set K of example (1), which consists of two arithmetic progressions with the same difference, characterizes in sufficient detail the more general situation of the case $T = 3k - 3$. In fact, the following theorem is true:

THEOREM. *Let $T = 3k - 3$. Then the following cases are possible*:
1. *The length of the set K does not exceed $2k - 1$.*
2. *K is isomorphic to K_0.*
3. *$k = 6$ and K is isomorphic to the set*

$$K_6 = \{(0,0), (0,1), (0,2), (1,0), (1,1), (2,0)\}.$$

Thus the set K is contained in a set of integers which is isomorphic to the set of interior lattice points of some convex closure of a domain D, with bounded volume in the euclidean space E_n, which is one-dimensional in case 1 and two-dimensional in cases 2 and 3. Now it is clear that Theorem 1.11 describes the structure of the set K in the same terms as Theorem 1.9, the only difference being that in Theorem 1.9 the domain D is one-dimensional.

In the example which we have considered all essential features of the structure of K for large values of T are also present.[3]

1.12. PROOF OF THEOREM 1.11 ON THE STRUCTURE OF THE SET K FOR $T = 3k - 3$. We shall prove the theorem by induction. Its validity for $2 \leqslant k \leqslant 5$ follows from Theorem 1.6. Assume that the theorem is true for all integers less than a fixed number k with $k \geqslant 6$. Then let

$$K = \{a_0, a_1, a_2, \ldots, a_{k-1}\}, \, a_i < a_{i+1}, \quad i = 0, 1, 2, \ldots, k - 2.$$

We may assume that $a_0 = 0$ and $d(k) = 1$. Then $a_{k-1} \geqslant 2k - 1$. We use the notation $K' = K \setminus a_{k-1}$ and $K'' = K \setminus a_0$.

First we assume that $d(K') = d_1 > 1$. Then the numbers $a_{k-1} + a_s, s = 0, 1, \ldots, k - 1$, are not contained in the set $2K'$. Therefore $T' = 2k - 3$, $K' = \{0, a, 2a, \ldots, (k-2)a\}$, $a > 1$, and K is isomorphic to K_0 with $k_1 = k - 1$. In the sequel we shall assume that $d_1 = 1$.

For the proof we distinguish between the following two cases:
A) $a_{k-2} \geqslant 2k - 3$ or $a_{k-1} - a_1 \geqslant 2k - 3$.
B) $a_{k-2} \leqslant 2k - 4$ and $a_{k-1} - a_1 \leqslant 2k - 4$. In the case A) we may restrict our attention to the case $a_{k-2} \geqslant 2k - 3$, since if the inequality $a_{k-1} - a_1 \geqslant 2k - 3$ is satisfied, then we may consider the set K^* (§1.10) instead of K. Finally, in this case we may restrict our attention to a set K with the following property:

A_2) $a_{k-2} \geqslant 2k - 3$, and among the numbers $a_{k-1} + a_s, s = 0, 1, \ldots, k - 1$, there are no more than three elements of the set $2K'$.

---

[3] The reader may now survey the content of the Fundamental Theorem given in §2.10.

We show that this is the case.

1) $a_{k-1} - a_{k-2} = 1$, $a_{k-1} - a_{k-3} > 2$,

2) $a_{k-1} - a_{k-2} \geqslant 2$.

These two cases were considered in the proof of Theorem 1.10.

3) $a_{k-1} - a_{k-2} = 1$ and $a_{k-1} - a_{k-3} = 2$. Considering the set K^* we obtain $a_1 = 1$ and $a_2 = 2$.

3a) $a_i < 2i$, $i = 1, 2, \ldots, k - 2$. If one of the numbers $2k - 3$ or $2k - 2$ were contained in the set $2K'$ then the argument of case 3a) in Theorem 1.10 would show that $T \geqslant 3k - 2$, which is impossible. Thus the numbers $2k - 3$ and $2k - 2$ do not belong to the set $2K'$. Then the numbers

$$2k - 3 - a_1, 2k - 3 - a_2, \ldots, 2k - 3 - a_{k-2} \tag{1.12.1}$$

and

$$2k - 2 - a_1, 2k - 2 - a_2, \ldots, 2k - 2 - a_{k-2} \tag{1.12.2}$$

are not contained in K'. There exist $k - 1$ integers not exceeding $2k - 3$ and not belonging to the set K'. But for the set (1) the number of elements is equal to $k - 2$, and thus the set (2) has at most one element different from the numbers (1). Hence the set (1) consists of $k - 2$ consecutive integers and $K' = \{0, 1, 2, \ldots, k - 2\}$.

Now it is evident that the numbers 0, 1 and a_{k-2} are not contained in the set $2K''$. We want to point out, however, that in the case considered above we have shown that K is isomorphic to K_0.

3b) $a_i < 2i$, $i = 1, 2, \ldots, k - 3$, and $a_{k-2} = 2k - 4$. In this case all the numbers from 0 to $2k - 6$ are contained in the set $2K'$; $a_{k-2} = 2k - 4$, $a_{k-2} + a_1 = 2k - 3$ and $a_{k-2} + a_2 = 2k - 2$; thus $T \geqslant 2k - 5 + 3 + k = 3k - 2$, which contradicts the assumption of the theorem. Hence this case is impossible.

3c) Suppose there exists a number j, $3 \leqslant j \leqslant k - 2$, such that

$$a_i < 2i, \quad i = 1, 2, \ldots, j - 1, \ a_j > 2j.$$

We consider the sets $\bar{K}$ and $\bar{\bar{K}}$ as defined in (1.10.4) and (1.10.5). It follows from Theorem 1.10 that $\bar{T} \geqslant 3j$. To the set $\bar{\bar{K}}$ we may also apply Theorem 1.10, since we may always assume that $\bar{\bar{K}}$ contains two numbers different from 1. This is the case if $a_{k-1} - a_{k-2} = 1$ or $a_{k-2} \leqslant 2k - 4$. If $a_{k-1} - a_{k-2} > 1$ and $a_{k-2} \geqslant 2k - 3$ then we have case 2). For $\bar{\bar{T}}$ we obtain the inequality $\bar{\bar{T}} \geqslant 3(k - j)$. Thus

$$T \geqslant \bar{T} + \bar{\bar{T}} - 3 \geqslant 3k - 3.$$

Since $T = 3k - 3$, we have $\bar{T} = 3j$. As we have shown under case 3a), this implies

$$\bar{K} = \{0, 1, 2, \ldots, j-1, a_j\}.$$

Thus the numbers 0, 1 and a_j are not contained in the set $2K''$.

3d) Suppose there exists a number j, $4 \leqslant j \leqslant k - 2$, such that

$$a_i < 2i, \; i = 1, 2, \ldots, j-2, \; a_{j-1} = 2j - 2, \; a_j > 2j.$$

The proof is carried out as in case 3c), where the inequality $\bar{T} \geqslant 3j + 1$ is obtained from 3b), and thus it follows that this case is impossible.

3e) $a_i < 2i$, $i = 1, 2, \ldots j - 2$, $a_{j-1} = 2j - 3$, and $a_j = 2j$, $4 \leqslant j \leqslant k - 2$. The proof proceeds as in case 3c), and the inequality $\bar{T} \geqslant 3j + 1$ results from the fact that the set $2K$ contains all the numbers from 0 to $2j - 1$ ($2j - 2 = a_{j-1} + a_1, 2j - 1 = a_{j-1} + a_2$) and $j + 1$ numbers of the form

$$a_j + a_s, \qquad s = 0, 1, \ldots, j.$$

3f) $a_i < 2i$, $i = 1, 2, \ldots, j - 3$, $a_{j-2} = 2j - 4$, $a_{j-1} \leqslant 2j - 2$, and $a_j > 2j$, $5 \leqslant j \leqslant k - 1$. It should be noted that if $a_1 = 1$ and $a_2 = 2$, then either 0, 1, a_3 do not belong to the set $2K''$, or a_3 must assume one of the two values 3 or 4. Therefore the integers from $2j - 4$ to $2j$, inclusively, do not belong to the set $2K$ and we obtain $\bar{T} \geqslant 3j + 1$ as above.

3g) $a_i < 2i$, $i = 1, 2, \ldots, j - 2$, $a_{j-1} = 2j - 2$ and $a_{j+1} = 2j$, $4 \leqslant j \leqslant k - 3$. As in case 3b) we consider the sets $\bar{K}$ and $\bar{\bar{K}}$. The number $a_{j-2} + a_{j+1}$ belongs neither to $d\bar{K}$ nor to $2\bar{\bar{K}}$. Therefore $T > \bar{T} + \bar{\bar{T}} - 3 + 1 \geqslant 3k - 2$.

3h) $a_i < 2i$, $i = 1, 2, \ldots, j - 2$, $a_{j-2} \leqslant 2j - 6$, $a_{j-1} = 2j - 2$, and $2j + 1 \leqslant a_{j+1} \leqslant 2j + 2$, $4 \leqslant j \leqslant k - 3$. Let us consider the sets

$$\bar{K}_1 = \{0, a_1, \ldots, a_{j-1}, a_{j+1}\}$$

and $\bar{\bar{K}}$. The general elements of the sets $2\bar{K}_1$ and $2\bar{\bar{K}}$ are $2a_{j-1}, a_{j-1} + a_{j+1}$ and $2a_{j+1}$ (the number $a_{j-1} + a_j$ cannot belong to the set $2\bar{K}_1$ inasmuch as $a_{j-1} + a_j > a_{j-2} + a_{j+1}$.) Therefore (if at least one of the numbers $2j - 3$, $2j - 2$ belongs to the set $2\bar{K}_1$) we have

$$T \geqslant \bar{T}_1 + \bar{\bar{T}} - 3 \geqslant 3j + 1 + 3(k - j) - 3 \geqslant 3k - 2.$$

We show that we have thus exhausted all possible cases. If condition 3a) is not satisfied then there exists a number j for which

(I) $a_i < 2i$, $i \leqslant j - 1$.

(II) $a_j \geqslant 2j$.

(III) $j \leqslant k - 2$.

Under 3c) the case $a_j > 2j$ was considered, and the case $a_{k-2} = 2k - 4$ was treated under 3b). Thus there remain for consideration the conditions

(I) $a_i < 2i, i \leqslant j - 1$.

(II*) $a_j = 2j$.

(III*) $j \leqslant k - 3$.

instead of (I)–(III). After dealing with the case 3d) there remains (taking into account the difference in the notation of indices) the condition

(IV) $a_{j+2} \leqslant 2j + 2$.

After excluding the case 3e) one has to deal with the condition

(V) $a_{j-1} \leqslant 2j - 4$;

after case 3f) has been completed, there remains the condition

(VI) $a_{j+2} \leqslant 2j + 4$.

After considering condition 3g) the remaining case is

(VII) $a_{j+2} \geqslant 2j + 3$.

Case 3e) completes the argument.

Thus we have shown that in the case A) we may restrict our consideration to sets K with the property A_2). It is easy to verify the conditions of the theorem for the set K'. Since $d(K) = 1$ and $a_{k-2} \geqslant 2k - 3$, Theorem 1.10 implies $T' \geqslant 3k - 6$. But in view of A_2) one has $T' \leqslant T - 3 = 3k - 6$, and thus $T' = 3k - 6$.

Thus we may apply the induction hypothesis to the set K'. If K' is not isomorphic to K_6, then K' has the form

$$K' = \{0, a, 2a, \dots, (k_1 - 1)a, b, b + a, \dots, b + (k_2 - 1)a\}, \qquad (1.12.3)$$

where $k_1 + k_2 = k - 1$, and a and b are positive integers.

We now take up the excluded case where $k = 7$ and K' is isomorphic to K_6. Then one of the points $(0, 0)$, $(0, 2)$ or $(2, 0)$ corresponds to the number 0.

In each of these cases the set K' is expressible in the form

$$K' = \{0, b, 2b, a, a + b, 2a\}, \qquad b < a, (a, b) = 1.$$

Then

$$
2K' = \{\; 0 \quad\quad b \quad\quad 2b \quad\quad 3b \quad\quad 4b
$$
$$
a \quad\quad a + b \quad\quad a + 2b \quad\quad a + 3b
$$
$$
2a \quad\quad 2a + b \quad\quad 2a + 2b
$$
$$
3a \quad\quad 3a + b
$$
$$
4a\}. \tag{1.12.4}
$$

Inasmuch as $a_5 + a_6$ and $2a_6$ do not belong to the set $2K'$, one of the numbers a_6 and $a_6 + a$ must be an element of this set.

a) $a_6 + a_4 = 2a_5$, i.e. $a_6 = 3a - b$. Comparing the numbers $3a - b$ and $4a - b$ with the numbers (4), we find that always $a \leqslant 5$, which is impossible since $a_5 = 2a \geqslant 11$.

Thus, if we have, for example, $4a - b = 2b$ and $4a = 5b$ then $a = 5$.

b) $a_6 + a_4 \neq 2a_5$. In this case $a_6 + 2b \in 2K'$. By equating this number with the different numbers from (4) we obtain the following cases: $a_6 + 2b = 3a$, $a_6 + b = 3a - b \in 2K'$, which was discussed under a); furthermore,

$$a_6 + 2b = 3a + b, \ a_6 = 3a - b \in 2K'$$

$$a_6 + 2b = 4a, \ a_6 + b = 4a - b \in 2K'.$$

Thus we have shown that the set K' is not isomorphic to K_6, and therefore it must have the form (3).

We show that in this case K is isomorphic to K_0, where we shall not use the condition $a_{k-1} \geqslant 2k - 1$ as given, thus allowing for the case $a_{k-1} = 2k - 2$.

First we assume that $a = 1$. In this case

$$K' = K_1' \cup K_2', \quad \text{where } K_1' = \{0, 1, \dots, k_1 - 1\},$$

$$K_2' = \{b, b + 1, \dots, b + k_2 - 1\}.$$

If $a_{k-1} = b + k_2 + 1$ then

$$T = T(2K_1') + T(K_1' + \{K_2', a_{k-1}\}) + T(2\{K_2', a_{k-1}\}) \geqslant 3k - 2. \quad (1.12.5)$$

Now let $a_{k-1} > b + k_2 + 1$. If $k_2 = 1$, then $2a_{k-1}$, $a_{k-1} + a_{k-2}$ and two of the three numbers $a_{k-1} + a_s$ with $s = k - 5$, $k - 4$, $k - 3$ do not belong to the set $2K'$; hence

$$T \geqslant T' + 4 = 3k - 2. \quad (1.12.6)$$

Let $k_2 = 2$. Suppose that $k = 6$ and that the numbers

$$a_{k-1} + a_s \quad (1.12.7)$$

for $s = 0, 1, 2$ coincide with the numbers of the set $2K_2'$. Then we obtain $a_{k-1} = 2b$ and K turns out to be isomorphic to K_6. In the opposite case one of the numbers (7) with $k = 6$ and $s = 0, 1, 2$ or one of the numbers (7) with $k - 7 \leqslant s \leqslant k - 4$ is not contained in the set $2K'$. Since the numbers (7) with $k - 3 \leqslant s \leqslant k - 1$ are not contained in the set $2K'$ either, the relation (6) must hold.

Finally, if $k_2 \geqslant 3$ then the numbers (7) with $k - 4 \leqslant s \leqslant k - 1$ do not belong to the set $2K'$, which again implies (6).

Thus, having eliminated the special case where $a_{k-1} = 2b$ with $k = 6$ and $k_2 = 2$, we always have $a_{k-1} = b + k_2$.

In the case $a = 2$, if $a_{k-1} = a_{k-2} + 4$, the inequality (5) is obtained. The further argument is carried out in analogy to the case $a = 1$.

We show that for $a > 3$ it is impossible to have simultaneously the congruences

$$a_{k-1} \not\equiv 0 \ (\text{mod } a) \quad \text{and} \quad a_{k-1} \not\equiv b \ (\text{mod } a).$$

First let $a = 3$. Then we must have $a_{k-1} \equiv 2b \ (\text{mod } a)$. The set K' is of the form

$$K' = \{0, 3, \ldots, 3(k_1 - 1), b, b + 3, \ldots, b + 3(k_2 - 1)\}.$$

Let us assume at first that the number $3(k_1 - 1)$ is the largest element of the set K'. Then $3(k_1 - 1) > 2k - 4$ and thus $k_1 > (2k - 1)/3$. The numbers

$$a_{k-1}, a_{k-1} + 3, \ldots, a_{k-1} + 3(k_1 - 1) \tag{1.12.8}$$

must belong to the following numbers from the set $2K'$:

$$2b, 2b + 3, \ldots, 2b + 3(2k_2 - 2). \tag{1.12.9}$$

These are $2k_2 - 1$ numbers. Since $k_1 + k_2 = k - 1$, it follows that

$$2k_2 - 1 = 2k - 2k_1 - 3 < k_1 - 2.$$

Therefore there are no more than three among the numbers (8) which do not belong to the set $2K'$, and if we also take into account the number $2a_{k-1}$, we obtain $T \geqslant T' + 4 \geqslant 3k - 2$, which is impossible.

Now suppose that the largest number in the set K' is $b + 3(k_2 - 1)$.

If $k_2 = 1$ then two of the numbers $a_{k-1} + a_s$, $s = k - 5, k - 4, k - 3$, do not belong to the set $2K'$.

If $k_2 \geqslant 2$ and $3(k_1 - 1) < b + 3(k_2 - 2)$ the proof proceeds as in the case $a = 1$. Now let $3(k_1 - 1) > b + 3(k_2 - 2)$, $k_2 \geqslant 2$. If $b \equiv 1 \ (\text{mod } 3)$ then it is impossible that $a_{k-1} - a_{k-2} > 1$, inasmuch as one then has $a_{k-1} - a_{k-2} \geqslant 4$, and the numbers $a_{k-1} + a_s$, $s = k - 4, k - 3, k - 2, k - 1$, cannot belong to the set $2K'$.

Let us consider the case $a_{k-1} - a_{k-2} = 1$. Since $b + 3(k_2 - 1) > 2k - 4$, it follows that $3(k_1 - 1) > 2k - 5$ and $k_1 > (2k - 2)/3$.

Therefore $2k_2 < 1$. Thus there are not more than two among the numbers (8) which are not contained in the set $2K'$, and, furthermore, the integers $a_{k-1} + a_{k-2}$ and $2a_{k-1}$ are not among these numbers.

If $b \equiv 2 \ (\text{mod } 3)$ then we have $a_{k-1} - a_{k-2} = 2$, and the numbers

$$a_{k-1} + b + 3(k_2 - 1), \ a_{k-1} + b + 3(k_2 - 2)$$

do not belong to the set $2K'$. Since $b + 3(k_2 - 1) > 2k - 4$, one has

$$3(k_1 - 1) > 2k - 6 \quad \text{and} \quad k_1 > (2k - 3)/3,$$

which implies $2k_2 - 1 < k_1$. At least one of the numbers (8) does not belong to the set $2K'$, and together with $2a_{k-1}$ there are always at least four of these numbers.

Now let $a \geqslant 4$. In this case the numbers of the form

$$a_{k-1} + sa, \qquad s = 0, 1, \ldots, k_1 - 1,$$

can belong to the set $2K'$ only if they are of the form $a_j + a_t$, where $a_j, a_t \equiv b$ (mod a), and the numbers of the form

$$a_{k-1} + b + sa, \qquad s = 0, 1, \ldots, k_2 - 1,$$

may coincide with numbers from the set $2K'$ only if they are of the form $a_j + a_t$, where $a_j, a_t \equiv 0$ (mod a). Two numbers, one of them of the form $a_{k-1} + s_1 a$, the other of the form $a_{k-1} + b + s_2 a$, cannot both be elements of the set $2K'$, since then the two congruences $a_{k-1} \equiv 2b$ (mod a) and $a_{k-1} + b \equiv 0$ (mod a) would hold simultaneously, and this would imply $3b \equiv 0$ (mod a). But this is impossible for $a \geqslant 4$ in view of $(a, b) = 1$.

Suppose one of the two congruences mentioned is true, say $a_{k-1} \equiv 2b$ (mod a). In this case, for any number a_j with $a_j \equiv b$ (mod a), the number $a_{k-1} + a_j$ does not belong to the set $2K'$. The number of such integers is k_2. Among the k_1 numbers of the form $a_{k-1} + a_j$ with $a_j \equiv 0$ (mod a) no less than $\max(0, k_1 - (2k_2 - 1))$ are contained in the set $2K'$.

The general number of elements of the form $a_{k-1} + a_s, s = 0, 1, \ldots, k - 1$, which do not belong to the set $2K'$ is not less than $1 + k_2 + \max(0, k_1 - 2k_2 + 1)$. This number is never less than 4 (after excluding the case where it is equal to 3 with $k_2 = 2$ and $k_1 = 3$, which leads to the case where K is isomorphic to K_6).

The case $a_{k-1} \equiv -b$ (mod a) is treated analogously.

Thus we have shown that for $a \geqslant 3$ one of the congruences $a_{k-1} \equiv 0$ (mod a) or $a_{k-1} \equiv b$ (mod a) is satisfied.

Suppose that the second one of these congruences holds. Then the numbers

$$a_{k-1} + b + (k_2 - 1)a, \quad a_{k-1} + (k_1 - 1)a, \quad 2a_{k-1}$$

do not belong to the set $2K'$. Therefore, if $k_1 = 1$, then

$$a_{k-1} + b + (k_2 - 2)a = 2b + 2(k_2 - 1)a,$$

and if $k_1 \geqslant 2$ then

$$a_{k-1} + (k_1 - 2)a = (k_1 - 1)a + b + (k_2 - 1)a,$$

i.e. $a_{k-1} = b + k_2 a$. The case

$$a_{k-1} \equiv 0 \pmod{a}$$

is also handled analogously.

The case A) has thus been completed. We proceed to consider case B).

It is evident that $a_1 > 2$. We determine the number $j \geqslant 2$ from the condition $a_i > 2i$ for $i = 1, 2, \ldots, j - 1$, $a_j \leqslant 2j$. Then, obviously,

$$a_{j-1} = 2j - 1, \, a_j = 2j.$$

We consider the sets $\bar{K}$ and $\bar{\bar{K}}$ as defined in (1.10.4) and (1.10.5). As in case 3c) we can show here that $\bar{T} = 3j$ and $\bar{\bar{T}} = 3(k - j)$.

We show that $a_{j-2} < 2j - 2$. Suppose that $a_{j-2} = 2j - 2$ and consider instead of $\bar{\bar{K}}$ the set

$$\bar{\bar{K}}_1 = \{a_{j-2}, a_{j-1}, \ldots, a_{k-1}\}.$$

Then $\bar{\bar{T}}_1 \geqslant 3(k - j) + 3$, and since the sets $2\bar{K}$ and $2\bar{\bar{K}}_1$ have the five common points $2a_{j-2} + s$, $s = 0, 1, 2, 3, 4$, this implies

$$T \geqslant \bar{T} + \bar{\bar{T}}_1 - 5 \geqslant 3k - 2,$$

which contradicts the assumption of the theorem.

In exactly the same way we may show that $a_{j+1} > 2j + 1$.

Since the numbers $a_j + a_s$, $s = j - 2, j - 1, j$, are not contained in the set $2K'$, we have

$$\bar{T}' = 3j - 3.$$

In exactly the same way one can prove the equation $\bar{\bar{T}}'' = 3(k - j) - 3$.

By applying the induction hypothesis we may conclude that the sets $\bar{K}'$ and $\bar{\bar{K}}''$ each consist of two arithmetic progressions. If we apply the argument of case A) and take into account that the stipulation $a_{k-1} \geqslant 2k - 1$ has not been used anywhere, we find that the sets $\bar{K}$ and $\bar{\bar{K}}$ are also of this structure.

a) $j = 2$, $a_1 = 3$ and $a_2 = 4$. In this case $\bar{\bar{K}} = K''$. If the number a in K'' is equal to 3 or 4 then the proof is completed. If $a = 5$ then $a_3 = 8$ and $a_4 = 9$ or 13. Thus $0, a_1, a_2$ and a_4 are not contained in the set $2K''$. If $a > 5$ then $0, a_1, a_2$ and a_3 do not belong to $2K''$.

b) $j = k - 2$. It is evident that, because $a_{j-2} = 2j - 3$, the number a in $\bar{K}$ must equal 3. The inequality $a_{k-1} \geqslant a_{k-2} + 4$ is impossible, since then the numbers $a_{k-1} + a_s$, $s = k - 4, k - 3, k - 2, k - 1$, would not belong to the set $2K'$. Therefore $a_{k-1} = a_{k-2} + 3$.

c) $2 < j < k - 2$. In view of $a_{j-2} = 2j - 3$ the number a in $\bar{K}$ must equal 3. If in the set $\bar{K}$ the number a is equal to 3 then the proof is completed. Suppose that $a = 4$. Then $a_{j+1} = a_{j-1} + 4$, $a_{j+1} = a_j + 4$ or $a_{j+2} = a_{j-1} + 8$. In any case the number $a_{j-2} + a_{j+2}$ belongs neither to $2\bar{K}$ nor to $2\bar{\bar{K}}$. Then $T \geqslant \bar{T} + \bar{\bar{T}} - 3 + 1 \geqslant 3k - 2$. But if we assume that $a \geqslant 5$ then the number $a_{j-2} + a_{j+1}$ does not belong to either $2\bar{K}$ or to $2\bar{\bar{K}}$. Consequently $a = 3$.

We have completed B) and thus finished the proof of the theorem.

EXERCISE 1. Show by means of an example that the result of Theorem 1.11 is best possible in case 1.

1.13. *The structure of K for $T = 3k - 2$.* With growing T the study of the structure of K becomes altogether more complicated, as one might observe for example, from the proof of Theorems 1.9 and 1.11.

In the paper [10] the case $T = 3k - 2$ was treated by elementary methods. The following theorem was proved:

THEOREM. *Let $t = 3k - 2$. Then the following cases are possible*:
1. *The length of the set K does not exceed $2k + 1$.*
2. *K is isomorphic to the set*

$$K_0 = \{(0,0), (0,1), \ldots, (0, k - 3), (0, k - 1), (1, 0)\}.$$

3. *For $k = 10$ the set K may be isomorphic to*

$$K_{10} = \{(0,0), (0,1), (0,2), (0,3), (1,0), (1,1), (1,2), (2,0), (2,1), (3,0)\}.$$

The proof of this theorem is very tedious and it shall be omitted here.

For investigating the structure of the set K in the case $T > 3k - 2$ it turns out to be necessary to use analytic methods. These results will be presented in Chapter II.

EXERCISE 1. Show by means of an example that the result of Theorem 1.13 is best possible in case 1.

1.14. *A lower bound for T in the case $K \subset E_n$.*

NOTATION. Let E_n be the euclidean space of n dimensions. An *integral vector* shall be understood to be a vector with integer coordinates. Sometimes a vector shall also be called a *point* and shall be denoted by a letter with a bar over it or in the form $(x_1, \ldots, x_n)$, where the x_i, $1 \leqslant i \leqslant n$, are the coordinates of the vector.

Z_n is the additive group of integral vectors in E_n. $\{\bar{e}_j\}$, $1 \leqslant i \leqslant n$, is an orthonormal basis in E_n.

Cosets of a linear subspace of dimension s, $1 \leqslant s \leqslant n - 1$, shall be called *planes*. In particular, we obtain *straight lines* for $s = 1$ and *hyperplanes* for $s = n - 1$.

The dimension of a point set is defined as the dimension of the set of all vectors which join any pair of points from this set. The linear subspace generated by the vectors $\bar{a}, \bar{b}, \ldots, \bar{l}$ is denoted by $L(\bar{a}, \bar{b}, \ldots, \bar{l})$.

LEMMA. *Let $K \subset E_n$ be a finite set of dimension n with $T(K) = k$. Then*

$$T \geqslant (n + 1)k - n(n + 1)/2. \tag{1.14.1}$$

The proof of the inequality (1) will proceed by induction. The lemma is true for $n = 1$ with arbitrary k. For any n and $k = n + 1$ the inequality (1) is valid, since then $T = (n + 1)(n + 2)/2$. Now let us assume that (1) has been proved for $n = 1, 2, \ldots, m - 1$ with any k and for $n = m$, $k = m + 1, m + 2, \ldots, p - 1$ (since K has dimension n, one has $k \geqslant n + 1$.) Then we show the validity of (1) for $n = m$, $k = p$.

We consider the smallest convex polyhedron containing the set K and having among its vertices the point $\bar{a}$ which obviously belongs to the set K.

Suppose $K \setminus \{\bar{a}\}$ is contained in some hyperplane. The set $K \setminus \{\bar{a}\}$ has dimension $n - 1$; otherwise, if the dimension were smaller, then the dimension of K could not be larger than $n - 1$. Therefore $T(2(K \setminus \{\bar{a}\})) \geqslant m(p - 1) - (m - 1)m/2$ and

$$T = T(2(K \setminus \{\bar{a}\})) + T((K \setminus \{\bar{a}\}) + \{\bar{a}\}) + 1$$

$$\geqslant m(p - 1) - (m - 1)m/2 + p - 1 + 1$$

$$= (m + 1)p - m(m + 1)/2.$$

Now suppose the set $K \setminus \{\bar{a}\}$ has dimension m. We consider the smallest convex polyhedron containing $K \setminus \{\bar{a}\}$. Let L be its face relative to which the point $\bar{a}$ lies on the same side as some points of $K \setminus \{\bar{a}\}$. Then L contains at least m points from K. Therefore

$$T \geqslant T(2(K \setminus \bar{a})) + T(K \cap L + \bar{a}) + 1$$

$$\geqslant (m + 1)(p - 1) - m(m + 1)/2 + m + 1$$

$$= (m + 1)p - m(m + 1)/2.$$

1.15. *The structure of $K \subset Z_2$ with $T < 10k/3 - 5$.* In §1.13 we discussed the difficulty of investigating the structure of the set K for $T > 3k - 3$ by elementary methods. The solution of this problem becomes simpler if, instead of a set of integers, one considers a set of lattice points (assumed, of course, to be noncollinear) in the plane. In §§1.15–1.17 we shall investigate the structure of the set $K \subset Z_2$ with

$$3k - 3 \leqslant T < 10k/3 - 5. \tag{1.15.1}$$

LEMMA. *Let* $K \subset Z_2$ *and* $T < 10k/3 - 5$. *Then the following cases are possible*:
1. *The set* K *lies on two parallel straight lines.*
2. *The set* K *with* $k = 10$ *is isomorphic to* K_{10}.

First we consider the case where the set K is decomposable into subsets $K_1, \ldots, K_s$ which lie on s parallel lines. Then

$$
\begin{aligned}
T &\geqslant T(2K_1) + T(K_1 + K_2) + T(2K_2) \\
&\quad + T(K_2 + K_3) + \ldots + T(K_{s-1} + K_s) + T(2K_s) \\
&\geqslant 2k_1 - 1 + k_1 + k_2 - 1 + 2k_2 - 1 + \ldots + 2k_s - 1 \\
&= 4k - (k_1 + k_s) - (2s - 1)
\end{aligned}
\tag{1.15.2}
$$

and also

$$
\begin{aligned}
T &\geqslant T(2K_1) + T(K_1 + K_2) + T(K_1 + K_3) + \ldots + T(K_1 + K_s) \\
&\quad + T(K_2 + K_s) + \ldots + T(2K_s) \\
&\geqslant 2k_1 - 1 + k_1 + k_2 - 1 + k_1 + k_3 - 1 + \ldots + k_1 + k_s - 1 \\
&\quad + k_2 + k_s - 1 + \ldots + 2k_s - 1 \\
&= 2k + (k_1 + k_s)(s - 1) - (2s - 1).
\end{aligned}
\tag{1.15.3}
$$

It follows from (2) and (3) that

$$T \geqslant (4 - 2/s)k - (2s - 1). \tag{1.15.4}$$

If $3 \leqslant s \leqslant k/3$, then (4) implies that $T \geqslant 10k/3 - 5$, which proves the lemma in the case $s \leqslant k/3$.

The remainder of the proof will proceed by induction. For $k \leqslant 6$ we have from (1) the inequality $T \leqslant 3k - 4$. Then it follows from Theorem 1.9 that K is situated on one side of a straight line. If $7 \leqslant k \leqslant 9$ then (1) implies $T \leqslant 3k - 3$. Then it follows from Theorem 1.11 that the set K lies on two straight lines. If $10 \leqslant k \leqslant 12$ then (1) implies $T \leqslant 3k - 2$. In this case it follows from Theorem 1.13 that the set K lies on two straight lines, excepting the possibility of the set K being isomorphic to K_{10} in the case $k = 10$.

We assume that the assertion of the lemma has been proved for all values not exceeding $k - 1$, where $k \geqslant 13$, and we show that it is then also valid for k.

We consider the convex hull D of the set K, i.e. the smallest convex set containing K. Obviously D is a polygon all of whose vertices are contained in K. We consider an arbitrary vertex $\bar{a}_1$ as well as the two edges of the boundary of D which intersect at $\bar{a}_1$, and the two additional vertices $\bar{a}_2$ and $\bar{a}_3$ on these edges. Let us consider the lattice generated by the points $\bar{a}_1$, $\bar{a}_2$ and $\bar{a}_3$. An arbitrary vector $\bar{b}$ from the set K is uniquely expressible in the form $\bar{b} = \alpha_1(\bar{a}_2 - \bar{a}_1) - \alpha_2(\bar{a}_2 - \bar{a}_1)$. Among the vectors $\bar{b}$ for which the numbers α_1 and α_2 are not both integers, we select one in such a way that there does not exist any vector $\bar{b}' \neq \bar{b}$ with coefficients α_1', α_2' satisfying $\alpha_1' \leqslant \alpha_1$ and $\alpha_2' \leqslant \alpha_2$.

The points $2\bar{a}_1$, $\bar{a}_1 + \bar{a}_2$, $\bar{a}_1 + \bar{a}_3$ and $\bar{a}_1 + \bar{b}$ do not belong to the set $2(K \setminus \{\bar{a}_1\})$. Therefore $T(2(K \setminus \{\bar{a}_1\})) \leqslant T - 4 < 10(k - 1)/3 - 5$, and the set $K \setminus \{\bar{a}_1\}$ is located on two parallel lines. Thus the set K lies on no more than three lines. Since $s < k/3$, this proves the lemma.

Thus we may assume that all α_1, α_2 are pairs of integers. We consider the linear transformation of the plane for which $\bar{a}_1 \to (0,0)$, $\bar{a}_2 \to (0,1)$ and $\bar{a}_3 \to (1,0)$. The set K is then mapped onto an isomorphic set which lies in the first quadrant where the points have nonnegative integer coordinates. As before, we shall denote the image of K by K again.

Let us suppose that there are no more than five points from the set K on the line $x_1 = 0$. We consider s lines parallel to $x_1 = 0$ on which the points of the set K are located. Then we only have to consider the case $S > k/3$. In view of (3) we have

$$T \geqslant 2k + (s - 1)(k_1 + k_s - 2) - 1 > 2k + (k/3 - 1)(6 - 2) - 1$$
$$= 10k/3 - 5,$$

which is impossible in view of condition (1).

Thus there are no more than four points on the line $x_1 = 0$. If the number of these points is greater than two, then $(0,2) \in K$. Indeed, if $t > 1$ is the smallest of the numbers for which $(0,t) \in K$, then $(0,0) + (0,0)$, $(0,0) + (1,0)$, $(0,0) + (0,1)$, $(0,0) + (0,t) \notin 2(K \setminus \{(0,0)\})$. If there are four points on the coordinate axis then $(0,3) \in K$. Otherwise it may be mapped onto $(0,0)$ by a suitable linear transformation, and one would only have to repeat the argument just outlined.

Suppose that $(0,c)$ and $(1,d)$ are the points with maximal ordinates on the lines $x_1 = 0$ and $x_1 = 1$, respectively. If $d \leqslant c$ then the set K lies on no more than four lines, which proves the lemma. It remains to consider the case $d > c$. We use the notation $K_1 = K \cap \{x_1 = 0\}$, $K_2 = K \cap \{x_1 = 1\}$. Then

$$T(K_1 + K_2) \geqslant T(K_1 + \{1,0\}) + T(K_1 + \{1,d\}) = 2T(K_1).$$

Furthermore, $T(2K_1) \geqslant 2T(K_1) - 1$. Thus

$$T(2(K \setminus K_1)) < 10k/3 - 5 - 4T(K_1) + 1 < (10/3)(k - T(K_1)) - 5.$$

If $T(K_1) = 3$ and $k = 13$, then $T(2(K \setminus K_1)) \leqslant 27$. If $T(K_1) = 4$ and $k = 14$, then $T(2(K \setminus K_1)) \leqslant 26$. Hence the set $K \setminus K_1$ lies on two parallel lines. On the line $x_1 = 1$ there are at least 3 points from the set K. Otherwise, $(0, 0) + (1, d)$ $\notin 2(K \setminus \{(0, 0)\})$. Consequently the lines which contain the set $K \setminus K_1$ must be parallel to the line $x_1 = 0$, and hence K lies on no more than three parallel lines. This proves the lemma.

EXERCISES. 1. For a set $K \subset Z_2$ which is not contained in any single straight line, the inequality $T \geqslant 3k - 3$ holds.

2. Generalize Lemma 1.15 to the case of three lines.

3. Prove that, if $K \subset Z_2$ and $T < (4 - 2/s)k - (2s - 1)$ and if k is sufficiently large, then there exist $s - 1$ parallel lines which cover K.

4**. Let $K \subset Z_2$ be a set not containing any three collinear points. Find a lower bound for T.

5**. Generalization of the preceding exercise: Let $K \subset Z_n$ be a set not containing any m points within the same s-dimensional hyperplane. Find a lower bound for T.

We may consider the special cases $s = 1$ (straight line) and $s = n - 1$ (hyperplane) for $n = 2$ and other small values of n, for $m = s + 2$ and other small values of $m > s + 2$; furthermore, the case $m = [k^\varepsilon]$, $0 < \varepsilon < 1$.

1.16. *Projection of a set of lattice points onto a plane.* We consider all points from a given set $K \subset E_n$ of integral vectors for which the first $n - 1$ coordinates $x_1, \ldots, x_{n-1}$ have equal values. For some fixed $(n - 1)$-tuple $x_1, \ldots, x_{n-1}$ let the number of such points in K be s, i.e. points of the form $\bar{x}_i = (x_1, \ldots, x_{n-1}, x_{n_i})$, $1 \leqslant i \leqslant s$. Instead of these s points we choose the points $(x_1, \ldots, x_{n-1}, u)$, $u = 0$, $1, \ldots, s - 1$. This process is performed for all fixed $x_1, \ldots, x_{n-1}$ with $s \geqslant 1$. The set K^0 so obtained is called the projection of the set K onto the plane

$$L(\bar{e}_1, \bar{e}_2, \ldots, \bar{e}_{n-1})$$

parallel to $\bar{e}_n$. If $s \leqslant 1$ for any $(n - 1)$-tuple $x_1, \ldots, x_{n-1}$, we obtain the ordinary projection.

THEOREM. *If K^0 is the projection of the set K onto the plane $L(\bar{e}_1, \ldots, \bar{e}_{n-1})$ parallel to $\bar{e}_n$, then*

$$T(2K^0) \leqslant T(2K). \tag{1.16.1}$$

PROOF. We consider numbers $b_1, \ldots, b_{n-1}$ such that the set K contains two points

$$\bar{x}' = (x_1', x_2', \ldots, x_n') \quad \text{and} \quad \bar{x}'' = (x_1'', x_2'', \ldots, x_n'')$$

which satisfy the relations

$$x'_j + x''_j = b_j, \qquad 1 \leqslant j \leqslant n - 1. \tag{1.16.2}$$

Suppose in the set K^0 the values of s determined by the numbers $x'_1, \ldots, x'_{n-1}$ and by $x''_1, \ldots, x''_{n-1}$, respectively, are s_1 and s_2. Let $t = \max(s_1 + s_2 - 1)$, where the maximum is taken over all pairs $\bar{x}', \bar{x}''$ with the property (2). For given $b_j, 1 \leqslant j \leqslant n - 1$, the points

$$(b_1, b_2, \ldots, b_{n-1}, u), \qquad u = 0, 1, \ldots, t - 1$$

and only these are contained in the set $2K^0$. But the set $2K$ contains at least t points whose first $n - 1$ coordinates have given values $b_1, \ldots, b_{n-1}$.

REMARK. Let $\bar{a}_1, \ldots, \bar{a}_n$ be a given basis of Z_n. Then it is evident how the projection of K onto the hyperplane $L(\bar{a}_1, \ldots, \bar{a}_{n-1})$ is to be defined in the respective coordinate system. Property (1) is preserved.

1.17. *Precise structure of $K \subset Z_2$ for $T < 10k/3 - 5$.*

THEOREM. *Let $K \subset Z_2$ be a set which cannot be embedded in any straight line, with $T < 10k/3 - 5, k \geqslant 11$.[4] Then K is contained in a set which is isomorphic to*

$$K_0 = \{(0,0), (0,1), \ldots, (0, k_1 - 1), (1,0), (1,1), \ldots, (1, k_2 - 1)\},$$

where $k_1, k_2 \geqslant 1, k_1 + k_2 = T - 2k + 3$.

PROOF. It follows from Lemma 1.15 that the set K lies on two parallel lines l_1 and l_2. We may assume that K lies on the lines $x_2 = 0$ and $x_2 = 1$, since there does always exist a linear transformation of the plane which maps the lattice points of the lines l_1 and l_2 onto lattice points of the lines $x_2 = 0$ and $x_2 = 1$, and the set K is therefore isomorphic to this set.

Let the set of abscissae for $x_2 = 0$ and $x_2 = 1$, respectively, be equal to $\{a_0, \ldots, a_{m-1}\}$ and $\{b_0, \ldots, b_{n-1}\}$, $m + n = k$. We may assume that $a_0 = 0, b_0 = a_{m-1}$ and $(a_1, \ldots, a_{m-1}, b_1, \ldots, b_{n-1}) = 1$; otherwise there would exist an inverse image of the set K satisfying these conditions. We project the set K onto the line $x_2 = 0$, thus obtaining a set $K^0 = K_1^0 \cup K_2^0$, where the set K_1^0 is such that the ordinates of its points are all equal to zero while the set of abscissae is $\{a_0, \ldots, a_{m-1}, b_1, \ldots, b_{n-1}\}$, and the set K_2^0 consists of the single point $(b_0, 1)$.

Since by Theorem 1.16

$$T(2K^0) = T(2K_1^0) + T(K_1^0 + K_2^0) + T(2K_2^0) = T(2K_1^0) + k - 1 + 1$$

$$\leqslant T(2K),$$

---

[4] For $k < 11$ the structure of the set K was obtained in the proof of Lemma 1.15.

one has

$$T(2K_1^0) \leqslant T - k,$$

and therefore an application of Theorem 1.9 to the set K_1^0 yields

$$b_{n-1} - b_0 + a_{m-1} - a_0 = k_1 + k_2 - 2 \leqslant T - 2k + 2.$$

EXERCISES. 1. Give an example to show that Theorem 1.17 cannot be sharpened by reducing the quantity $k_1 + k_2$.

2. Give an example to show that Theorem 1.17 cannot be sharpened by increasing the upper bound for T.

3. Generalize the example of the exercises in §§1.11 and 1.12 by giving examples of sets K of length $k + b$ for which $T = 2k - 1 + b$ and which cannot be mapped isomorphically onto the set K_0 of §1.17.

4. Formulate hypotheses which are analogous to Theorems 1.11 and 1.13.

5. Formulate and prove a sharpening of Theorem 1.17 by giving the maximum possible value of the upper bound for T under the additional assumption that K lies on two straight lines.

§3. The function $W(r, T, k)$

1.18. *Definition of the function* $W(r, T, k)$. In this section we define and study a function which describes in sufficient detail the structure of the set K for given T.[5]

An isomorphic mapping of $K \subset E_m$ into E_n shall be called *regular* if the image of K has dimension n, and *singular* if the image of K has a dimension less than n.

Thus the mapping $K_{2,3} \to E_2 : 0 \to (0,0), 1 \to (1,1), 3 \to (3,3)$ is singular. The isomorphic image $K_{2,3} \to E_2 : 0 \to (0,0), 1 \to (1,0), 3 \to (0,1)$ is regular.

Suppose there exists a regular isomorphic image of the set K in E_n but there does not exist a regular isomorphic image of the set K in E_{n+1}. The number N with this property will be denoted by r. In this manner r is determined uniquely by the set K.

For example, $r(K_{1,3}) = 1$ since any isomorphic image of $K_{1,3}$ in E_n with $n \geqslant 2$ lies on a straight line. The values of r for the set considered in Theorem 1.6 are listed in Table 1. Figures Figures 1a and 1b show regular isomorphic images of these sets under mappings into E_r.

By V_K we denote the number of integral points in the closure of the set K. We introduce the notation

$$W_K = \min_{\varphi} V_{K\varphi}, \tag{1.18.1}$$

---

[5] The reader will undoubtedly notice the connection between the following exposition and the preliminary considerations in §1.11, which will be discussed in greater detail further below (§1.26).

where the minimum is taken with respect to the images of K, denoted by $K\varphi$, under all regular isomorphic mappings φ into Z_r of the given set K. Furthermore, we let

$$W(r, T, k) = \max_K W_K, \tag{1.18.2}$$

where the maximum is taken with respect to all sets K with given values r, T and k. The quantity W_K is a generalization of the concept of "length" of the set K as introduced in §1.11. The number W_K will be called the *volume* of the set K of lattice points.

EXERCISES. 1. Verify the values of r for the sets of Theorem 1.6 as given in Table 1.

2. Calculate the values of $W(r, T, k)$ for the arguments listed in Table 1.

3. Show that $r = 1$ if $T \leqslant 3k - 4$,

4. Show that

$$W(1, T, k) = T - k + 1$$

if $k \geqslant 3$ and $T \leqslant 3k - 4$,

5. Show that

a) $W(1, 3k - 3, k) = 2k - 1, k \geqslant 5$,

b) $W(2, 3k - 3, k) = k, k \geqslant 3$,

c) $W(1, 3k - 2, k) = 2k + 1, k \geqslant 10$,

d) $W(2, 3k - 2, k) = k + 1, k \geqslant 10$,

e) $W(2, T, k) = T - 2k + 3, T < 4k - 6$.

1.19. *On the domain of definition of the function* $W(r, T, k)$. The number k may assume arbitrary positive integer values. For T the inequality

$$2k - 1 \leqslant T \leqslant k(k + 1)/2$$

was shown in §1.8. In §1.14 we proved the inequality

$$T \geqslant (r + 1)k - r(r + 1)/2,$$

which gives an upper bound for r. For any k and arbitrary T and r satisfying the inequalities given above, there exists a set K with these values. Examples of such K are given in §§1.20 and 1.21.

1.20. *A lower bound for* $W(1, T, k)$. In this subsection we give examples of sets K with $r = 1$ for which T assumes values from $2k - 1$ to $(k - 1)k/2 + 2$. The values of W_K for these examples furnish a lower bound for $W(1, T, k)$.

For any given positive integer values s and t satisfying the inequalities $2 \leqslant s \leqslant k - 1, 1 \leqslant t \leqslant \max(1, k - s - 1)$, we consider the set

$$K(k, s, t) = \{0, 1, 2, \ldots, k - s, k - s + t,$$

$$2(k - s + t), 2^2(k - s + t), \ldots, 2^{s-2}(k - s + t)\}. \tag{1.20.1}$$

Computing the value $T(2K)$, we obtain

$$T = sk - s(s + 1)/2 + t + 1. \tag{1.20.2}$$

If s assumes values from 2 to $k - 1$, and t (for given s) runs through the values from 1 to $k - s - 1$ (for $s = k - 1$ the value of t is equal to 1), then T assumes values from $2k - 1$ to $(k - 1)k/2 + 2$. Conversely, if T is given such that

$$2k - 1 \leqslant T \leqslant (k - 1)k/2 + 2,$$

then s and t are determined by (2) by the formulas

$$s = \left[\frac{2k^{(1)} - 1 - ((2k^{(1)} - 1)^2 - 8T^{(1)} + 16)^{1/2}}{2} \right], \tag{1.20.3}$$

$$t = T^{(1)} - sk^{(1)} + s(s + 1)/2 - 1, \tag{1.20.4}$$

where $T^{(1)} = T$, $k^{(1)} = k$, and the number s in (4) is defined by (3).

A consideration of $K(k, s, t)$ gives the following lower bound for $W(1, T, k)$.

THEOREM. *For values of T satisfying*

$$2k - 1 \leqslant T \leqslant (k - 1)k/2 + 2,$$

the inequality

$$W(1, T, k) \geqslant 2^{s-2}(k - s + t) + 1 \tag{1.20.5}$$

holds with s and t defined by equations (3) *and* (4).

An easily verified estimate for the lower bound in (5) for small values of T is given at the end of §1.21 by the relation (21.4) with $r = 1$.

1.21. *A lower bound for $W(r, T, k)$.* For given values r, T and k (see 1.19) a set $P(r, T, k) \subset E_r$ is constructed as follows: We include in $P(r, T, k)$ the lattice points $(0, x_2, x_3, \ldots, x_r)$ for which $x_i \geqslant 0$ and $x_2 + x_3 + \ldots + x_r = 1$ (in other words, one of the coordinates x_i equals one and the others with $i \geqslant 2$ vanish). We also include in $P(r, T, k)$ the points $(x_1, 0, 0, \ldots, 0)$ with $\{x_1\} = K(k - r + 1, s, t)$, where s and t are defined by the equations (20.3) and (20.4) with

$$T^{(1)} = T - (2k - r + 2)(r - 1)/2, \tag{1.21.1}$$

and

$$k^{(1)} = k - r + 1. \tag{1.21.2}$$

It is easy to verify that $T = T(2P(r, T, k))$. It is also evident that $P(1, T, k) = K(k, s, t)$. An investigation of the set $P(r, T, k)$ yields the following lower bound for $W(r, T, k)$:

THEOREM. *For values of T with*

$$(r + 1)k - r(r + 1)/2 \leqslant T \leqslant (k - 1)k/2 + r + 1,$$

the inequality

$$W(r, T, k) \geqslant 2^{s-2}(k^{(1)} - s + t) + r, \tag{1.21.3}$$

is valid with s and t defined by (20.3) *and* (20.4) *in which $T^{(1)}$ ana $k^{(1)}$ have the values determined by* (1) *and* (2).

Let $\varepsilon > 0$ be a fixed number which may be chosen arbitrarily small. Then there exists a sufficiently small $c > 0$, depending on ε, such that for $T < ck^{3/2}$ the relations (20.3) and (1) imply the equation

$$s = [T^{(1)}/k^{(1)} + O(T^{(1)^2}/k^3)] = [T/k - r + 1 + \theta_1],$$

where θ_1 may be made arbitrarily small by a suitable choice of c; furthermore,

$$(1 - \theta_1)k < k^{(1)} - s + t < 2k$$

and

$$2^{s-2}(k^{-(1)} - s + t) + r = \theta \cdot 2^{1/k-r}k, \tag{1.21.4}$$

with

$$1/2 - \varepsilon < \theta < 2 + \varepsilon.$$

1.22. *A lower bound for $W(r, T, k)$ which is independent of T.*

THEOREM.

$$W(r, T, k) \leqslant 2^{k-1-r} + r. \tag{1.22.1}$$

PROOF. We may assume that the set K has dimension r and $K \subset E_r$. Let $K = \{\bar{a}_0, \ldots, \bar{a}_{k-1}\}$, where the lattice points $\bar{a}_i$, $0 \leqslant i \leqslant k - 1$, are ordered in the following way: $\bar{a}' = \{a'_1, \ldots, a'_r\}$ is "later" then $\bar{a}'' = \{a''_1, \ldots, a''_r\}$ if there exists an index s such that $a'_j = a''_j$, $1 \leqslant j \leqslant s - 1$, and $a'_s > a''_s$.

We construct an isomorphic mapping φ of the set K into E_{2k-2}. In order to accomplish this we proceed by induction, defining successively isomorphic mappings φ_{s-1} of the sets $K_s = \{\bar{a}_0, \ldots, \bar{a}_{s-1}\}$, $2 \leqslant s \leqslant k$, into E_{2k-2}.

It is immediately clear that the construction may be carried out in such a way that the first r coordinates of the vectors $\bar{a}_i$ and $\bar{a}_i \varphi_{s-1}$, $2 \leqslant s \leqslant k$, coincide for any fixed i, $0 \leqslant i \leqslant s - 1$. Thus any two distinct differences between elements from K_s are transformed into equal ones under the mapping φ_{s-1}. In order to prove that the mapping is an isomorphism we merely have to verify that equal differences are transformed into equal ones (criterion 1.5).

It should be noted, however, that among the differences $\pm(\bar{a}_s - \bar{a}_i)$, $1 \leqslant i \leqslant s - 1$, there are no identical ones, due to the method of ordering.

We let

$$\bar{a}_0 \varphi_1 = (a_{01}, a_{02}, \ldots, a_{0r}, 0, \ldots, 0),$$

$$\bar{a}_1 \varphi_1 = (a_{11}, a_{12}, \ldots, a_{1r}, 1, 0, \ldots, 0).$$

Suppose the image $K_s \varphi_{s-1}$ of the set K_s has been constructed and its dimension is equal to p_s. Then we construct the image $K_{s+1} \varphi_s$ of the set K_{s+1}. Here the following three cases are possible:

a) $\bar{a}_s - \bar{a}_i \neq \bar{a}_j - \bar{a}_t$, where i, j, t are arbitrary integers satisfying the condition $0 \leqslant i, j, t \leqslant s - 1$. In this case we have $K_s \varphi_s = K_s \varphi_{s-1}$ and

$$\bar{a}_s \varphi_s = (a_{s1}, a_{s2}, \ldots, a_{sr}, 0, 0, \ldots, 1, \ldots, 0),$$

where the number 1 occupies the $(p_s + r + 1)$th position. Then the preceding remarks imply immediately that the sets K_{s+1} and $K_{s+1} \varphi_s$ are isomorphic.

b) There exists a unique number t for which there is a pair i, j with $1 \leqslant i, j, t \leqslant s - 1$ satisfying

$$\bar{a}_s + \bar{a}_t = \bar{a}_i + \bar{a}_j. \tag{1.22.2}$$

In this case we obtain $K_s \varphi_s = K_s \varphi_{s-1}$ and

$$\bar{a}_s \varphi_s = \bar{a}_i \varphi_s + \bar{a}_j \varphi_s - \bar{a}_t \varphi_s. \tag{1.22.3}$$

The sets K_{s+1} and $K_{s+1} \varphi_s$ are isomorphic. In fact, there exists an isomorphism between the sets K_s and $K_s \varphi_s$

An equation of the form $\bar{a}_s - \bar{a}_p = \bar{a}_q - \bar{a}_u$, $0 \leqslant p, q, u \leqslant s - 1$, may hold only in the case where $u = t$ and $\bar{a}_p + \bar{a}_q = \bar{a}_i + \bar{a}_j$, which implies

$$\bar{a}_p \varphi_s + \bar{a}_q \varphi_s = \bar{a}_i \varphi_s + \bar{a}_j \varphi_s,$$

and therefore by (3) we have

$$\bar{a}_s \varphi_s - \bar{a}_p \varphi_s = \bar{a}_q \varphi_s - \bar{a}_t \varphi_s .$$

c) There exist more than one value of t for which there are pairs i, j satisfying p equations of the form (2), i.e.

$$\bar{a}_s + \bar{a}_{t_w} = \bar{a}_{i_w} + \bar{a}_{j_w}, \qquad 1 \leqslant w \leqslant p.$$

The points $\bar{a}'_{sw}$ are defined by equations of the form (3), i.e.

$$\bar{a}'_{sw} = \bar{a}_{i_w} \varphi_{s-1} + \bar{a}_{j_w} \varphi_{s-1} - \bar{a}_{t_w} \varphi_{s-1} . \tag{1.22.4}$$

Suppose these points (not all of which are different) determine a plane P_1 containing them whose dimension m is equal to the dimension of the set of these points.

The plane P_2 of dimension p_s which is parallel to $L(\bar{e}_{r+1}, \ldots, \bar{e}_{r+p_s})$ and runs through the point $(a_{s1}, \ldots, a_{sr}, 0, \ldots, 0)$, contains the plane P_1. Let $\bar{u}_1, \ldots, \bar{u}_m$ be a basis for the lattice consisting of all integral points of the plane P_1 and let this basis be enlarged by the vectors $\bar{w}_1, \ldots, \bar{w}_{p_s-m}$ to become a basis of the lattice $P_2 \cap Z_{2k-2}$. Let $\bar{x}\varphi_{pr}$ be the projection of the point $\bar{x} \in L(\bar{e}_1, \ldots, \bar{e}_{p_s+r})$ onto the plane $L(\bar{e}_1, \ldots, \bar{e}_r, \bar{w}_1, \ldots, w_{p_s-m})$ parallel to $L(\bar{u}_1, \ldots, \bar{u}_m)$. Furthermore, let

$$\bar{\varphi} : L(\bar{e}_1, \ldots, \bar{e}_r, \bar{w}_1, \ldots, \bar{w}_{p_s-m}) \to E_{2k-2}$$

be the homomorphism induced by the relations

$$\bar{e}_i \bar{\varphi} = \bar{e}_i, \qquad 1 \leqslant i \leqslant r, \qquad \bar{w}_j \bar{\varphi} = \bar{e}_{r+j}, \qquad 1 \leqslant j \leqslant p_s - m.$$

Then

$$K_s \varphi_s = K_s \varphi_{s-1} \varphi_{pr} \bar{\varphi}, \quad \bar{a}_s \varphi_s = \bar{s}'_{s1} \varphi_{pr} \bar{\varphi} \qquad (\bar{a}_{sw} \varphi_{pr} = \bar{a}'_{s1} \varphi_{pr} \text{ for any } w)$$

The sets K_{s+1} and $K_{s+1} \varphi_s$ are isomorphic, as we shall now show. If two differences in K_s are equal, then the corresponding differences in $K_s \varphi_{s-1}$ are also equal, and equal differences are mapped onto equal ones under a projection. Thus K_s is isomorphic to $K_s \varphi_{s-1} \varphi_{pr} = K_s \varphi_s$.

Now suppose the equality

$$\bar{a}_s - \bar{a}_{i_w} = \bar{a}_{j_w} - \bar{a}_{t_w}, \qquad 1 \leqslant i_w, j_w, t_w \leqslant s - 1$$

is true. In this case we consider the equation (4) for this particular value of w. After performing the projection we obtain

$$\bar{a}_s \varphi_s - \bar{a}_{i_w} \varphi_s = \bar{a}_{j_w} \varphi_s - \bar{a}_{t_w} \varphi_s . \tag{1.22.5}$$

For any p and q satisfying

$$\bar{a}_s - \bar{a}_p = \bar{a}_q - \bar{a}_{t_w},$$

we obtain the equation

$$\bar{a}_s \varphi_s - \bar{a}_p \varphi_s = \bar{a}_q \varphi_s - \bar{a}_t \varphi_s$$

as we did at the end of part b).

Each of the sets $K_s \varphi_{s-1}$ is trivially contained in the lattice generated by its elements. We will show by induction that the ratio of the volume of the convex hull of the set $K_s \varphi_{s-1}$ to the volume of a fundamental parallelpiped of this lattice (the so-called reduced volume) is bounded from above by the number $2^{s-1-p_s} / p_s!$. For $s = 2$ this ratio equals 1, and $p_s = 1$.

In case a) we obtained a p_{s+1}-dimensional pyramid, where $p_{s+1} = p_s + 1$, with height one and base of area not exceeding $2^{s-1-p_s} / p_s!$; hence the volume of this pyramid is not larger than

$$\frac{1}{(p_s + 1)!} 2^{s-p_s-1} = \frac{1}{p_{s+1}!} 2^{(s+1)-1-p_{s+1}}.$$

In case b) we obtain $p_{s+1} = p_s$.

Suppose there exists a convex polyhedron $D \subset E_n$ and four points $\bar{x}_i$ such that $\bar{x}_1, \bar{x}_2, \bar{x}_3 \in D$, $\bar{x}_4 \notin D$ and $\bar{x}_1 - \bar{x}_2 = \bar{x}_3 - \bar{x}_4 = \bar{e}$. Let D_1 be the convex hull of the set $D \cup \{\bar{x}_4\}$. Then $V(D_1) \leqslant 2V(D)$. In fact, in order to verify this inequality it is only necessary to observe that the distance between $\bar{x}_4$ and any edge of D containing interior points of D_1 is less than the maximum of the distances from $\bar{x}_1$ and $\bar{x}_2$ to that edge. (The first of these distances is not larger and the second not less than $|(\bar{e}_0, \bar{e})|$, where $\bar{e}_0$ is the perpendicular of length one on the edge under consideration.) This consideration yields the inequality $V(K_{s+1} \varphi_s) \leqslant 2V(K_s \varphi_s)$.

Finally, in case c) we choose $m + 1$ points of the form (4) which span the plane P_1. The convex hull D_2 of the set consisting of the convex hull of $K_s \varphi_{s-1}$ and the given $m + 1$ points has a reduced volume not exceeding

$$\frac{1}{p_s!} 2^{s-1-p_s} \cdot 2^{m+1} = \frac{1}{p_s!} 2^{s+m-p_s}.$$

If a convex set $D \subset E_n$ with volume $V(D)$ is given and h is the distance between the supporting hyperplanes orthogonal to $\bar{e}_n$, then the projection of D onto $L(\bar{e}_1, \ldots, \bar{e}_{n-1})$ has a volume V_1 for which

$$V_1 \leqq nV(D)/h$$

(the maximal volume of the basis for a given volume of the solid D and given height h is attained by the pyramid).

The projection parallel to $L(\bar{u}_1, \ldots, \bar{u}_m)$ corresponds to successive projection parallel to $\bar{u}_1, \ldots, \bar{u}_m$, where the system of these vectors may be chosen in such a way that each of them may be embedded within D_2. As a result we obtain a reduced volume not exceeding

$$\frac{p_s(p_s - 1) \ldots (p_s - m + 1)}{p_s!} 2^{s+m-p_s} = \frac{1}{p_{s+1}!} 2^{(s+1)-1-p_{s+1}},$$

using the relation $p_{s+1} = p_{s-m}$. Consequently the reduced volume of the set $K\varphi$ is not greater than

$$\frac{1}{r!} 2^{k-1-r}.$$

A basis of the lattice generating $K\varphi$ may be mapped onto an orthonormal basis E_r, and thus we obtain an isomorphic image of the set K whose convex hull has a volume not exceeding

$$\frac{1}{r!} 2^{k-1-r}.$$

Now we show that this volume cannot contain more than $2^{k-1-r} + r$ lattice points from Z_r, provided, of course, that the set of these points has dimension r.

If the volume is equal to $1/r!$ and if it contains $r + 1$ lattice points such that the dimension of this set is equal to r, then the parallelepiped generated by these points must be a fundamental domain; thus the tetrahedron generated by these points must have the volume $1/r!$ and it cannot contain additional lattice points.

Each polyhedron with its vertices located at lattice points may be decomposed into tetrahedra with volume $1/r!$ such that any two of them are either disjoint or have a face in common. This fact is easily established by double induction with respect to the dimension r and the number k of lattice points within the polyhedron.

Suppose the polyhedron has volume $t/r!$ and may therefore be decomposed into t tetrahedra. Then we may order these tetrahedra in such a way that each of them abuts with one of the preceding ones. In the first tetrahedron there are $r + 1$ lattice points, and each of the subsequent ones adds at most one lattice point. Hence the total number of lattice points is not greater than $r + t$. In the present case we have $t \leqslant 2^{k-1-r}$, which proves the theorem.

EXERCISES. 1) Show that $W(r, (k - 1)k/2 + r + 1, k) = 2^{k-r-1} + r$.

2) Compute the value of $W(r, T, k)$ for large T.

1.23. *Estimation of $W(r, T, k)$ for $T < Ck$.* The estimate (22.1) is very weak. It suffices to compare it with the lower bound $W(r, T, k) \geqslant k$, which is obviously best possible. The equation $W(r, T, k) = k$ holds, for example, if $r = 1$ and $T = 2k - 1$.

In the case $T < Ck$, where C is an arbitrary given positive constant independent of k, we shall obtain in Chapter II a lower bound of $W(r, T, k)$ for sufficiently large k. The asymptotic order of this bound will be best possible. Indeed, the following theorem is true:

THEOREM. *For $T < Ck$, $C \geqslant 2$, there exist positive constants k_0 and c, depending only on C, such that $W(r, T, k) < ck$ for $k > k_0$.*

This result is the basic one in the present book. Its proof will occupy the whole of Chapter II.

1.24. *A conjecture on the function $W(r, T, k)$.* One may conjecture that the inequality (21.3) cannot be sharpened; in other words, that the following theorem is true.

THEOREM. *The function $W(r, T, k)$ is the one defined in the relation (21.3) with inequality replaced by the equality sign, i.e.*

$$W(r, T, k) = 2^{s-2}(k^{(1)} - s + t) + r, \qquad (1.24.1)$$

where

$$s = \left[\frac{2k^{(1)} - 1 - ((2k^{(1)} - 1)^2 - 8T^{(1)} + 16)^{1/2}}{2} \right],$$

$$t = T^{(1)} = sk^{(1)} + \frac{s(s + 1)}{2} - 1,$$

$$T^{(1)} = T - \frac{(2k - r + 2)(r - 1)}{2},$$

$$k^{(1)} = k - r + 1.$$

For $T < ck^{3/2}$, $c = c(\varepsilon) \geqslant 0$, the relations (1) and (21.4) imply

$$W(r, T, k) = \theta \cdot 2^{T/k-r}k,$$

where

$$1/2 - \varepsilon < \theta < 2 + \varepsilon, \qquad \varepsilon \geqslant 0.$$

Cases where one succeeds in determining the exact value of the function $W(r, T, k)$ have been studied in Exercises 2, 4 and 5 of §1.8 and Exercise 1 of §1.22. The answers to these exercises corroborate the conjecture that the theorem stated above be true in general.

Exercises. 1*. Determine $W(1, 3k - 2, k)$ without using Theorem 1.13.

2**. Determine $W(1, 3k - 1, k)$.

3***. Prove conjecture 1.24.

§4. Additive number theory as the theory of the invariants of isomorphisms

1.25. *An isomorphic mapping is an affine mapping.*

First we have to explain the meaning of the assertion made in the heading of this subsection, and also underline its limitations. We have to clarify whether it is possible to find an affine mapping which induces a given isomorphism. It is evident that for this version of the problem one has to consider two isomorphic subsets of the euclidean space E_n with an additive group of mappings.

At a first glance it seems as if the answer to this question would be negative. Indeed, the isomorphism between the two collinear sets $\{0, 1, 3\}$ and $\{0, 1, 4\}$ cannot be induced by any affine mapping. As still another example, consider the two planar sets $\{(0, 0), (1, 0), (3, 0)\}$. and $\{(0, 0), (1, 0), (0, 1)\}$.

Now we introduce the following restriction. We assume that the dimension of both isomorphic sets under consideration, as well as the dimension of the euclidean space, are equal to the size r of these sets. This condition suffices already to furnish an affirmative answer to the problem. We observe that in the first example all dimensions are equal to 1 but $r = 2$; in the second example the dimension of the first set is 1 but $r = 2$.

Now we are able to formulate and prove the relevant theorem.

Theorem. *Let two isomorphic sets of dimension r be given. More precisely, let*

$$K, K' \subset E_r, \quad K \sim K', \quad K = \{\bar{a}_0, \bar{a}_1, \ldots, \bar{a}_{k-1}\},$$

$$K' = \{\bar{a}'_0, \bar{a}'_1, \ldots, \bar{a}'_{k-1}\}, \quad \bar{a}_i \sim \bar{a}'_i, \bar{a}_0 = \bar{a}'_0 = 0.$$

Then there exist two bases of the lattice Z_r such that the coordinates of the points of K relative to the first basis are the same as the coordinates of the points of K' relative to the second basis.

Corollary. *Under the conditions of Theorem 1.25 there exists an additive mapping of the space E_r onto itself which induces an isomorphic mapping of the set K onto K'.*

Proof of the theorem. Theorem 1.25 follows almost immediately from the proof of Theorem 1.22. Indeed, for given s we choose from the system

$$\bar{a}_0 \varphi_{s-1}, \bar{a}_1 \varphi_{s-1}, \ldots, \bar{a}_{s-1} \varphi_{s-1}$$

of vectors a linearly independent subsystem in the following manner: We include the vector $\bar{a}_1 \varphi_{s-1}$ in this system. If the vectors $\bar{a}_{i_1} \varphi_{s-1}, \ldots, \bar{a}_{i_t} \varphi_{s-1}$ have already been chosen, then we select among the vectors $\bar{a}_p \varphi_{s-1}$ the vector with the smallest index $p > i_t$ which is linearly independent of these t vectors already chosen. The number of vectors in the linearly independent system so obtained is equal to p_s. The linearly independent system of r vectors obtained as the set $K\varphi$ at the last step may be projected onto $L(\bar{e}_1, \ldots, \bar{e}_r)$. The process of constructing the sets $K_s \varphi_{s-1}$ and of choosing a linearly independent subsystem is completely determined and unique for any two isomorphic sets K and K'. The process thus described leads to two bases of Z_r relative to which corresponding points of K and K' have the same coordinates.

Theorem 1.25 means that for two isomorphic sets K and K' satisfying the conditions of the theorem there exists a regular linear transformation of the space E_r onto itself which maps K onto K'.

1.26. *On the possibility of inducing an isomorphism by a homomorphism.* If we go through the argument of §1.22 under the assumption that the set $K \subset E_m$ has a dimension not exceeding r and that $\bar{a}_0 = 0$, we obtain analogously an isomorphic mapping φ of the set K into E_{k-1} for which $K\varphi$ has dimension r. Choosing a basis in $K\varphi$ (for the space E_r generated by $K\varphi$) as we did in §1.25, we obtain a homomorphic mapping $\varphi^{-1} : E_r \to E_m$ which induces an isomorphic mapping $\varphi^{-1} : K\varphi \to K$. Thus we have proved the following theorem:

THEOREM. *Let $K = \{\bar{a}_0, \ldots, \bar{a}_{k-1}\}$, $\bar{a}_0 = 0$, $K \subset Z_m$. Then there exists a homomorphism $\varphi : Z_r \to Z_m$ and a set $K_1 \subset Z_r$ of dimension r such that $K_1 \varphi = K$ and K_1 is isomorphic to K.*

Thus an isomorphism between two finite sets of vectors one of which has dimension r may induce a homomorphic mapping between vector spaces containing them.

Now we are in a position to consider in greater detail the question stated at the beginning of §1.18. There we remarked on the connection between results of the form mentioned in §1.11 and results which one obtains if the values of the function $W(r, T, k)$ are known.

In both cases we obtain a closed convex domain D of bounded volume in the euclidean space E_n. Let K' be the set of all lattice points in D. In the first case K is contained in the isomorphic image of the set K'; in the second case K is isomorphic to a certain subset K'' of K'. It is evident that in the first case, generally speaking, more information on the structure of the set K is given than in the second case. Thus, for example, the set $\{0, 2, 8\}$ is contained in a set which is isomorphic to the set $\{0, 1, 2, 3, 4\}$, i.e. in an arithmetic progression of length 5, but the set $\{0, 1, 100\}$, which may be mapped isomorphically into the set $\{0, 1, 2, 3, 4\}$, has length greater than 5. Within the scope of this subsection we shall say that in the first case a complete description of the structure of K is given, whereas in the second case only a partial description of the structure is provided. It turns out, if the dimension n of

D is equal to r, then a complete description of the structure of K may be derived from the partial description. In order to accomplish this, the first step is to apply Theorem 1.26. Suppose the set K is isomorphic to the set K_1 of dimension r, where the volume of the set of lattice points of K' contained in the closure of K_1 does not exceed $W(r, T, k)$. By Theorem 1.26 there exists a homomorphism $\varphi: Z_r \to Z_1$ such that $K'\varphi \supset K$, i.e. K is contained in the image K' under some homomorphism. In Chapter II we shall show how by means of a homomorphic mapping of the set K' of all lattice points of the set D one may obtain an isomorphic mapping (in the sense of §1.2) of the set of lattice points of some new domain D' whose image again contains K. This leads to the same description of the structure of K.

1.27. *A general point of view on additive problems.* In his "Erlanger Programm" Felix Klein proposed to consider geometry as the theory which studies the properties of figures which are invariant under all mappings from a given group of transformations. Isomorphic sets possess identical additive properties. Therefore we may formulate the following point of view on additive problems: Additive number theory is the theory which investigates those properties of sets of numbers which are invariant under isomorphic mappings.

By an *invariant* we mean a number which characterizes a property of a set and which remains unchanged under isomorphic mappings of that set. Thus we may say that additive number theory investigates those properties of sets of integers which are invariant under isomorphisms.

In order to make these questions specific we also have to give a precise description of the form of an isomorphic mapping (see §§1.3 and 1.4). It should be kept in mind that in our investigation an isomorphism is an isomorphism of order 2.

1.28. *Definition of the fundamental invariants.* Let $K \subset Z_n$. Since it is always possible to map the set K isomorphically into Z_1, we may assume without loss of generality that $K \subset Z_1$. As some of the fundamental invariants of the set K we define the following objects:

1) $T = T(2K)$ is that invariant which is basic for additive number theory.

2) The number R of different positive differences among the elements of K, i.e. the number of elements within the set

$$P = \{x, x = a_i - a_j, a_i > a_j\}.$$

3) Let $2K = \{b_1, \ldots, b_T\}$; furthermore, let r_s be the number of representations of the number b_s in the form $b_s = a_i + a_j$, $s = 1, 2, \ldots, T$, counting two representations as different even if they differ only in the order of the terms. Then let

$$M = \sum_{s=1}^{T} r_s^2.$$

4) Let $P = \{c_1, \ldots, c_R\}$; furthermore, let u_s be the number of representations of the number c_s in the form $c_s = a_i - a_j$, $a_i > a_j$, $s = 1, 2, \ldots, R$. Then we introduce

$$M' = \sum_{s=1}^{R} u_s^2 .$$

5) The quantity r as defined in §1.18 is an invariant.

6) We order the numbers $b_i \in 2K$ in such a manner that $r_i \geqslant r_{i+1}$, $i = 1, 2, \ldots$, $T - 1$. Then the numbers $r_1, \ldots, r_T$ form a system of invariants.

7) We order the numbers $c_i \in P$ in such a way that $u_i \geqslant u_{i+1}$, $i = 1, 2, \ldots$, $R - 1$. Then the numbers $u_1, \ldots, u_R$ form a system of invariants.

8) The quantity W_K as defined in §1.18 is an invariant.

EXERCISE**. Find a complete system of invariants of a set K.

§5. Analysis of the problems

1.29. *Direct and inverse problems.* In §1.7 we subdivided additive problems into direct and inverse problems. Using the definition of an invariant (see §1.27), we may say that a direct problem is a problem concerning the computation of invariants of a set of integers, whereas an inverse problem is concerned with finding sets with given invariants.

In the classical additive problems one usually investigates invariants of a specific set of integers. If the structure of the sets to be added is not defined in advance, we may consider a range of possible values of these invariants. Trivial examples of this kind will be found in Exercise 1 of §1.29; the reader is also referred to the more complicated Exercises 5 and 6 of §1.30.

In the present book only one inverse problem is investigated in detail: the structure of a set with given T. It turns out that the structure of the set K is sufficiently well determined if T is small. However, if T is large then the proposed methods cannot yield good results. Indeed, the size of a point set onto which the set K may be mapped isomorphically has an order of growth not less than $2^{T/k-r}$ if $T = O(k^{3/2})$ (see §1.21). With growing T this size increases rapidly, thus rapidly diminishing the information on the structure of K.

How can this difficulty be surmounted? First we may consider new concepts for describing the structure of K. Some remarks in this direction will be made in §1.33. Second, we may use additional information on the set K, i.e. in addition to T we may assume the values of some other invariants to be known and then solve the corresponding inverse problem.

The solution of inverse problems is, of course, possible also for systems of invariants which do not contain T.

Some possible lines of approach are illustrated by the simple examples considered in Exercises 1–4 of §1.31.

EXERCISES. 1. Find the range of values for the following invariants:
a) R,
b) M,
c) M',
d) r.

2. Determine the structure of K if
a) $M' = (k - 1)k(2k - 1)/6$; $M' = (k - 1)k/2$;
b) $R = k - 1$; $R = k(k - 1)/2$;
c) $T = 2k + 2$, $M' = (k - 2)(k - 1)(2k - 3)/6 + 3$;
d) $u_1 = k - 1$; $u_1 = k - 2$.

1.30. *Dependence between invariants.* In some cases geometric invariants are related by functional dependence (the radius and the circumference of a circle; the lengths of the sides and the area of a triangle). We may raise the analogous question of finding functional relations between number-theoretic invariants.

The dependence between invariants is not always a functional one. They may be related by inequalities (the relations between the sides of a triangle or between perimeter and the area of a convex figure in the plane).

Finding similar relations between number-theoretical invariants is essential for the correct formulation of inverse problems.

EXERCISES. 1. Determine M in the cases $T = 2k - 1$, $2k$, $k(k + 1)/2$.
2. Determine M if $T = 3k - 3$ and $r = 2$.
3. Show that $M' = (M - k^2)/2$.
4. Find the range of values for M in the cases $T = 2k + 1$, $2k + 2$, $2k + b$ with $b < k - 3$.
5*. Find the range of values of M for given T.
6*. Find the range of values of R for given T.

1.31. *An inverse problem analogous to the isoperimetric problem.* It is known that the perimeter p and the area S of a planar figure are related by the inequality $S < p^2/4\pi$. It is evident that the values p and S do not determine the form of the figure uniquely. However, if S assumes its maximal value for given p, i.e. $S = p^2/4\pi$, then the figure under consideration is uniquely determined as a circle with radius $p/4\pi$. If the value of S is very close to $p^2/4\pi$ then the figure is almost a circle. It is now completely clear how to state the problems analogous to the isoperimetric problem described above.

EXERCISES. 1**. Solve the "isoperimetric" problem for the invariants T and M.
2**. The same for T and R.

3**. The same for T and r.

4**. The same for T, r and M.

1.32. *Definition of a monomorphic image of subsets.* As we know, Theorem 1.24, even if proved completely, would give a great deal of information on the structure of K only for small values of $W(r, T, k)$, i.e. for small T. In §§1.32 and 1.33 we outline a possible approach to further research.

DEFINITION. We consider subsets B' and C' of the set B and C, respectively, with algebraic operations. A mapping of B' onto C' is called a *monomorphism* if it is one-to-one and if there exists a naturally induced mapping $2B' \to 2C'$.

Thus, if any element from $2B'$ has two different representations $b_1 + b_2 = b_3 + b_4$ as sums of elements from B', then the condition for the existence of the mapping $2B' \to 2C'$ implies $c_1 + c_2 = c_3 + c_4$.

As in §1.5 we obtain a criterion for a monomorphic mapping of sets of vectors in the euclidean space: Let B', $C' \subset E_n$. Then a given one-to-one mapping φ: $B' \to C'$ is a monomorphism if any equation of the form $\bar{b}_1 - \bar{b}_2 = \bar{b}_3 - \bar{b}_4$ implies $\bar{c}_1 - \bar{c}_2 = \bar{c}_3 - \bar{c}_4$, where $\bar{b}_i \in B'$, $\bar{c}_i \in C'$, $b_i\varphi = c_i$, $1 \leqslant i \leqslant 4$. For the sets which we obtain in Theorem 1.6 with $k = 3$ and $k = 4$ the possible monomorphic mappings are given by Figure 2.

1.33. *On the possibility of a precise description of the structure of K.* Perhaps it is possible to make the structure of K precise by finding a set K_1 which can be mapped monomorphically onto K and for which neither the quantity W_{K_1} as defined in (18.1) nor the quantity $r(K_1)$ is too large. The larger the volume of the convex set containing the set K_1 to be covered, the more precisely we can determine the pattern of the mutual order among the points of K_1, and hence of K.

It is undesirable to let the quantity $r(K_1)$ increase. This is clearly demonstrated by the fact that a set K for which $T = k(k + 1)/2$, $r = k - 1$ and $W_K = k$ can be mapped monomorphically onto any set consisting of k numbers, and hence no information whatsoever on the structure of the latter set is provided.

We define the function

$$r(w, T, k) = \max_K \min_{W_{K_1} \leqslant w} r(K_1).$$

Here the minimum is taken with respect to all sets K_1 which can be mapped monomorphically onto the set K and which satisfy the condition $W_{K_1} \leqslant w$, where w is an arbitrary fixed positive integer with $w \geqslant k$. The maximum is taken over all sets K with given T.

Upper bounds for the function $r(w, T, k)$ also furnish additional information on the structure of K with given T.

One may hope that the values of this function are not large even for small values of w, provided that T is not very large.

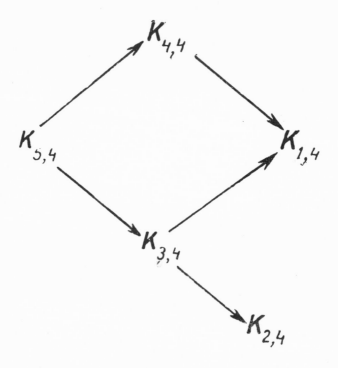

Figure 2

In order to have an example of a "good" covering we may consider a monomorphic mapping of the set

$$K_1 = \{0, \bar{e}_1, 2\bar{e}_1, \ldots, (k-s)\bar{e}_1, (k-s+t)\bar{e}_1, \bar{e}_2, \bar{e}_3, \ldots, \bar{e}_{s-1}\}$$

onto the set (see §1.20)

$$K(k,s,t) = \{0, 1, 2, \ldots, k-s, k-s+t,$$
$$2(k-s+t), 2^2(k-s+t), \ldots, 2^{s-2}(k-s+t)\},$$

where $2 \leqslant s \leqslant k-1$, $1 \leqslant t \leqslant \max(1, k-s-1)$. For K_1 the equations

$$W_{K_1} = k + t - 1, \qquad r(K_1) = s - 1$$

hold.

1.34. *Addition of some specific sets.* The results and propostions of the exercises in §§2–5 of this chapter may be successively generalized to the case where s identical sets are added.

By an *invariant of order* s we mean a number which characterizes a property of a set and which is invariant under isomorphic mappings of order s. We shall not give specific statements of problems analogous to those developed for the case $s = 2$; we mention, however, that new problems do arise. The exercises of §§1.30 and 1.31 may be considered as a system of invariants containing invariants of different order.

EXERCISES. 1*. Let $T_3 = T(3K)$. Find the range of values of T_3 for given T.

2*. Let $T_s = T(sK)$. Find the range of values of T_s for given T.

3*. Let $M_s = \int_0^1 |\sum_{j=0}^{k-1} e^{2\pi i \alpha a_j}|^{2s} d\alpha$. Find the range of values of M_s for given M. Show that $M_2 = M$.

1.35. *Addition of different sets.* The next step in generalizing the theory so far presented will be a study of the case where arbitrary sets are added without requiring them to be identical. In the paper [19] a generalization of Theorem 1.9 to the case of two different sets is given.

1.36. *Generalized form of the algebraic structure.* We have considered the basic case where $K \subset Z_n$. One may assume K to be a subset of any algebraic structure, for example of a nonabelian group.

CHAPTER II
THE FUNDAMENTAL THEOREM ON SUMS OF FINITE SETS

§1. The sum of sets of residue classes modulo a prime

2.1. *Introductory remarks.* In this chapter we develop a method which leads to a description of the structure of sets K of integers with small "double" sets $2K$ ($T < Ck, C \geqslant 2$). In the present section we give an illustration of the basic idea underlying the method by giving an example of an addition of sets of residue classes modulo a prime which is of interest in itself.

The problem of adding sets of residue classes was studied by Cauchy [9] and independently by H. Davenport [10]. They proved that if A and B are subsets of the ring S_p of residue classes modulo p, then $T(A + B) \geqslant \min (T(A) + T(B) - 1, p)$. This inequality is analogous to inequality (1.8.1).

In 1956 A. Vosper [32] studied the simplest inverse additive problem for sets of residue classes module a prime. He investigated the structure of A and B in the case when $T(A + B) = T(A) + T(B) - 1$ and he proved that if $T(A) + T(B) < p$ and $\min (T(A), T(B)) \geqslant 2$, then the sets A and B are arithmetic progressions in S_p with the same difference.

THEOREM. *Let $K = \{a_0, \dots, a_{k-1}\}$ be a set consisting of k elements from the group S_p of residue classes modulo p, where p is a prime. If $T < 12k/5 - 3$ and $k < p/35$, then the residue classes from K are contained in an arithmetical progression modulo p of length $k + b$, where b is defined by the equation $T = 2k - 1 + b$.*

For the proof of this theorem we apply a modified version of a method of trigonometric sums which is frequently used in number theory; furthermore we make use of the elementary results from §2 of Chapter 1, especially Theorem 1.9.

Before we turn to the proof of the theorem, we formulate and prove a lemma which will be the essential tool in proving the theorem.

2.2. *An inequality on the distribution of points on the circle.*

LEMMA.[1] *Let $\alpha_1, \dots, \alpha_k$ be numbers from the open interval $[0, 1)$. Let β be an arbitrary real number. Denote by $k_1(\beta)$ the number of indices $1, \dots, k$ for which the inequality*

$$\beta \leqslant \alpha_j < \beta + \tfrac{1}{2}(\mathrm{mod}\ 1)$$

---

[1] This lemma was formulated and proved in the paper[19]. Here we give a simpler proof due to L. P. Postnikova[37].

holds. i.e. for vhich there exists an integer m*, such that* $\beta + m \leqslant \alpha_j < \beta + \frac{1}{2} + m$. *If* $f(k)$ *is a function such that*

$$k_1(\beta) \leqslant f(k)$$

for any β, *then*

$$|S| \leqslant 2f(k) - k, \qquad where \ S = \sum_{j=0}^{k-1} e^{2\pi i \alpha_j}.$$

PROOF. Inasmuch as the equation

$$\left| \sum_{j=1}^{k} e^{2\pi i(\alpha_j + \gamma)} \right| = \left| \sum_{j=1}^{k} e^{2\pi i \alpha_j} \right|$$

holds for any real γ and the sequence $\{\alpha_j + \gamma\}$ also satisfies the inequality $k_1(\beta) \leqslant f(k)$, we may assume at once that

$$\left| \sum_{j=1}^{k} e^{2\pi i \alpha_j} \right| = \sum_{j=1}^{k} e^{2\pi i \alpha_j} = \sum_{j=1}^{k} \cos 2\pi \alpha_j.$$

Indeed,

$$\sum_{j=1}^{k} \cos 2\pi \alpha_j = \sum_{0 \leqslant \alpha_j < 1/4} \cos 2\pi \alpha_j - \sum_{1/4 \leqslant \alpha_j < 1/2} \cos 2\pi \left(\frac{1}{2} - \alpha_j \right)$$

$$- \sum_{1/2 \leqslant \alpha_j < 3/4} \cos 2\pi \left(\alpha_j - \frac{1}{2} \right) + \sum_{3/4 \leqslant \alpha_j < 1} \cos 2\pi (1 - \alpha_j)$$

$$= 2\pi \left(\sum_{0 \leqslant \alpha_j < 1/4} \int_{\alpha_j}^{1/4} \sin 2\pi t \, dt - \sum_{1/4 \leqslant \alpha_j < 1/2} \int_{1/2 - \alpha_j}^{1/4} \sin 2\pi t \, dt \right.$$

$$\left. - \sum_{1/2 \leqslant \alpha_j < 3/4} \int_{\alpha_j - 1/2}^{1/4} \sin 2\pi t \, dt + \sum_{3/4 \leqslant \alpha_j < 1} \int_{1 - \alpha_j}^{1/4} \sin 2\pi t \, dt \right)$$

$$= 2\pi \int_{0}^{1/4} \left(\sum_{0 \leqslant \alpha_j < t} 1 - \sum_{1/2 - t \leqslant \alpha_j < 1/2} 1 \right.$$

$$\left. - \sum_{1/2 \leqslant \alpha_j < 1/2 + t} 1 + \sum_{1 - t \leqslant \alpha_j < 1} 1 \right) \sin 2\pi t \, dt.$$

Since

$$k = 2\pi \int_0^{1/4} \left(\sum_{0 \leqslant \alpha_j < 1} 1 \right) \sin 2\pi t \, dt,$$

we have

$$\sum_{j=1}^k \cos 2\pi \alpha_j + k = 2\pi \int_0^{1/4} \left(\sum_{\substack{\alpha_j < 1/2 - t \\ \alpha_j \geqslant 1 - t}} 1 \right) \sin 2\pi t \, dt$$

$$+ 2\pi \int_0^{1/4} \left(\sum_{\substack{\alpha_j < t \\ \alpha_j \geqslant 1/2 + t}} 1 \right) \sin 2\pi t \, dt \leqslant 2f(k),$$

which proves the assertion.

COROLLARY. *If* $|S| > 2f(k) - k$ *then there exists a number* β *such that* $k_1(\beta) > f(k)$.

2.3. PROOF OF THEOREM 2.1. We consider the summation

$$I = \sum_{x_1, x_2 \in K} \sum_{x_3 \in 2K} \sum_{a=0}^{p-1} e^{2\pi i a (x_1 + x_2 - x_3)/p} = \sum_{a=0}^{p-1} S_1^2 S, \qquad (2.3.1)$$

where

$$S_1 = \sum_{j=0}^{k-1} e^{2\pi i (a/p) a_j}, \quad S = \sum_{x \in 2K} e^{-2\pi i (a/p) x}.$$

If the variable x runs through a complete system of residues modulo p, then

$$\sum_{x \in S_p} e^{2\pi i (a/p) x} = p \text{ if } p \text{ divides } a,$$

$$= 0 \text{ otherwise}$$

(see [40] for example), and one has

$$I = k^2 p. \qquad (2.3.2)$$

Indeed, for any two values x_1 and x_2 there exists one and only one value of x_3 for which $x_1 + x_2 - x_3 = 0$, and the number of such pairs is obviously equal to k^2.

If we assume that $|S_1| \leqslant 3k/5$ for any $a \not\equiv 0 \ (\mathrm{mod} \ p)$, then

$$|I| \leqslant k^2 T + \sum_{a=1}^{p-1} |S_1^2||S| \leqslant k^2 T + 3k/5k \left[\sum_{a=0}^{p-1} |S_1|^2 \sum_{a=0}^{p-1} |S|^2 \right]^{1/2}$$

$$= k^2 T + 3k/5\sqrt{kp \cdot Tp} \ .$$

Since $T < 12k/5$ and $k < p/35$, this implies $|I| < k^2p$, which contradicts the equation (2). Thus we have shown that there exists an $a' \not\equiv 0 \ (\mathrm{mod} \ p)$ such that

$$|S_1(a')| = \left| \sum_{j=0}^{k-1} e^{2\pi i(a'/p)a_j} \right| > 3k/5. \qquad (2.3.3)$$

If we analyze the argument just developed we observe immediately some peculiarities of applying the method of trigonometrical sums to the solution of inverse additive problems.

In order to solve direct additive problems by the method of trigonometric sums usually one applies the following scheme (see, for example, [40]).

A quantity to be determined (e.g. the number of representations of a number as a sum of terms with specified form) is expressed in the form of an integral (or summation) such that the integrand involves a trigonometric sum. The interval of integration (or the range of summation, respectively) is decomposed into two parts called major and minor intervals. The rest of the computation consists of the following sequence of steps:

a) One has to show that the absolute value of the trigonometric sum on the minor intervals possesses a good upper bound. This leads to an upper bound for the value of the integral on the minor intervals which turns out to be sufficiently small.

b) The part of the integral corresponding to the major intervals is transformed and furnishes the principal term of the ensuing asymptotic formula.

c) By combining the results obtained under a) and b) one shows that the value of the integral on the minor intervals is absorbed by the remainder term in the asymptotic formula for the integral on the major intervals. This also answers the question of giving an approximating expression for the quantity to be determined.

The solution of an inverse problem also begins by considering some integral (or summation) involving a trigonometric sum. In our example this is the summation I (1). Furthermore, as in the solution of a direct problem, we construct a decomposition into major and minor intervals. In the example given above the major intervals correspond to the part of the summation (1) for $a = 0$. The rest of the computation consists of the following sequence of steps which is reversed in comparison with the ordinary one:

a) The value of the integral (or summation) under consideration is easily calculated and thus known at the very beginning. In our example, $I = k^2 p$.

b) An upper bound for the value of the integral on the major intervals has to be established. This value turns out to be essentially smaller than the value of the integral on all intervals obtained under a). In our example one obtains the bound $k^2 T$ which does not exceed $7k^2 p/100$, since we have $T < 12k/5$ and $k < p/35$.

c) By combining the results of a) and b) one can show that the value of the integral on the minor intervals is large. This implies a lower bound for the absolute value of the trigonometric sum corresponding to some subset of the interval of integration; in the example under consideration this is the inequality (3).

Considering the system of inequalities thus obtained (in the present example it consists of one inequality only) one is able to obtain some original information on the structure of the set under discussion. We remark that the existence of a large absolute value for the trigonometric sum on the major intervals does not furnish any information on the structure of the set; thus, in our example, $S = k$ for $a = 0$ and any K.

The further argument contains three basic parts which may, under certain conditions, be described as follows:

1) "Definition of an essential part of the set K with maximum dimension."

2) "Definition of the compressed part of the set K."

3) "Contraction of the set K to the compressed part."

Step 1. It follows from the Corollary to Lemma 2.2 and from (3) that there exist integers u and v (the latter may be defined by means of the congruence $a'v \equiv 1 \pmod{p}$) such that the set K_1 of those residue classes in K which are congruent to the numbers

$$u + vs, \qquad 0 \leqslant s \leqslant (p - 1)/2, \qquad (2.3.4)$$

satisfies the condition

$$k_1 = T(K_1) \geqslant 4k/5. \qquad (2.3.5)$$

Suppose the residue classes from K_1 are obtained for the values of s contained in the set

$$P = \{s_0, s_1, \ldots, s_{k_1 - 1}\}, \qquad s_i < s_{i+1}, \qquad i = 0, 1, \ldots, k_1 - 2.$$

If necessary, we change the values of u and v in an unambiguous manner in such a way that the conditions $s_0 = 0$ and $d(P) = 1$ are satisfied. The set P of numbers is isomorphic to the set K_1. Indeed, if there are four residue classes of the form (4) such that the corresponding values $s^{(1)}$, $s^{(2)}$, $s^{(3)}$ and $s^{(4)}$ of s satisfy the congruence

$$u + vs^{(1)} + u + vs^{(2)} \equiv u + vs^{(3)} + u + vs^{(4)} \pmod{p},$$

then in view of (4) the relation

$$s^{(1)} + s^{(2)} \equiv s^{(3)} + s^{(4)} \pmod{p}$$

implies

$$s^{(1)} + s^{(2)} = s^{(3)} + s^{(4)}.$$

Thus different numbers from the set $2P$ cannot correspond to the same residue class from $2K_1$.

By means of (5) we have obtained the "essential" subset K_1 of the set K of residue classes, and we have shown that it is isomorphic to "a set of maximal dimension" which in this case is a set of integers.

Step 2. If the inequality $s_{k_1-1} \geqslant 2k_1 - 2$ were satisfied then Theorem 1.9 would imply

$$T \geqslant T(2K_1) \geqslant 3k_1 - 3 \geqslant 12k/5 - 3.$$

Thus $s_{k_1-1} \leqslant 2k_1 - 3$. We have shown that K_1 is contained in a short progression modulo p.

Step 3. If K contained a residue class a congruent to $u + vs$ with

$$4k_1 - 6 < s < p - (2k_1 - 3),$$

then the relation

$$T \geqslant T(2P) + T(P + a) \geqslant 2k_1 - 1 + k_1 = 3k_1 - 1$$

would follow. Thus all residue classes from K are congruent to numbers $u + vs$ with

$$-(2k_1 - 3) \leqslant s \leqslant 4k_1 - 6.$$

Hence the assertion of the theorem follows by subjecting K to the same argument which we applied to K_1.

EXERCISES. 1*. Sharpen Theorem 2.1 in the case $T < 3k - 3$.

2*. The condition $k < p/35$ is not necessary. Find a necessary and sufficient condition of this type under which Theorem 2.1 is true.

3*. Generalize Theorem 2.1 to the case of composite moduli.

§2. Some material from number theory

In this section we give a summary of number-theoretical results which will be used in the sequel.

2.4. *Farey sequences and Farey dissections*. The basic properties of Farey sequences can be found, for example, in the paper [40]. Here we will only give the necessary information without proofs.

Let us consider all reduced fractions between zero and one with positive denominators not exceeding the real number N, listing them in the order of increasing magnitude. This ordered set is called the *Farey sequence* of order N. Thus, for $N = 4.3$ the Farey sequence is

$$\frac{0}{1}, \frac{1}{4}, \frac{1}{3}, \frac{1}{2}, \frac{2}{3}, \frac{3}{4}, \frac{1}{1}.$$

If $p/1$ and p_1/q_1 are two consecutive fractions in a Farey sequence then

$$|p_1 q - p q_1| = 1.$$

If p/q and p_1/q_1 are two consecutive fractions in a Farey sequence then the (uniquely determined) mediant $(p + p_1)/(q + q_1)$ is an element of the Farey sequence of order $q + q_1$ which lies between them.

If p/q, p_1/q_1 and p_2/q_2 are three consecutive fractions of the Farey sequence of order N, then the interval

$$\left[\alpha'\left(\frac{p_1}{q_1}\right), \alpha''\left(\frac{p_1}{q_1}\right) \right] = \left[\frac{p + p_1}{q + q_1}, \frac{p_1 + p_2}{q_1 + q_2} \right)$$

containing the fraction p_1/q_1 is called *the interval of the fraction p_1/q_1*, and the system of these intervals for all fractions of the Farey sequence of order N is called the dissection of the interval $(0, 1)$ corresponding to the Farey sequence of order N, or more briefly, the *Farey dissection of order N*. The interval of a fraction p/q has a length not exceeding $2/qN$.

2.5. *Information from the geometry of numbers*.

THEOREM (MINKOWSKI). *Let $R \subset E_n$ be a convex domain which is symmetric with respect to the origin and has a volume $V > 2^n$ (possibly $V = \infty$). Then R contains a point with integer coordinates different from the origin.*

Proofs of this theorem and Theorem 2.6 are given in [7], pp. 150 and 154.

2.6. *Information from the geometry of numbers (continued)*. Let R be a convex, open domain which is symmetric with respect to the origin and has volume V, $0 < V < \infty$. For each I, $1 \leqslant I \leqslant n$, there exists a largest number λ, say λ_I, such that λR

contains I linearly independent points ([7], pp. 153–154). We shall call λ_I the Ith *successive minimum* of the domain R.

THEOREM (MINKOWSKI). *The successive minima satisfy the inequality*

$$V\lambda_1 \lambda_2 \ldots \lambda_n \leqslant 2^n.$$

§3. Formulation of the fundamental theorem

2.7. *The canonical parallelepiped.*

DEFINITION.[2] The parallelepiped $H = \{\bar{u} + \sigma_1 \bar{x}_1 + \ldots + \sigma_n \bar{x}_n\}$, where $\bar{u}, \bar{x}_i \in E_n$, $0 \leqslant \sigma_i < 1, 1 \leqslant i \leqslant n$, is called *canonical* if

$$(\bar{x}_1, \bar{x}_2, \ldots, \bar{x}_n) = (\bar{e}_1, \bar{e}_2, \ldots, \bar{e}_n) \begin{pmatrix} x_{11}x_{12}\ldots x_{1n} \\ x_{22}\ldots x_{2n} \\ 0 \cdot\cdot\cdot\cdot\cdot\cdot \\ \cdot\cdot\cdot\cdot\cdot\cdot \\ x_{nn} \end{pmatrix}, \qquad (2.7.1)$$

where $\bar{u}$ is an integral vector, the $h_j = x_{jj}$ are positive integers and

$$|x_{ij}| > x_{jj}, \qquad i < j, \qquad 1 \leqslant j \leqslant n. \qquad (2.7.2)$$

We observe that the number of lattice points within a canonical parallelepiped is equal to its volume.

In the case $n = 2$ a canonical parallelepiped is shown in Figure 3.

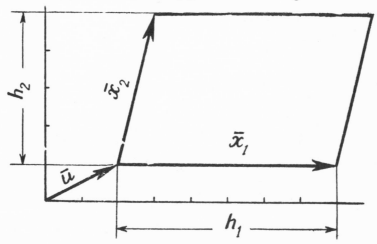

Figure 3.

---

[2] The definition given here differs somewhat from the original definition introduced in [21], p. 156. There $\bar{u} = 0$ and $h_j \geqslant 2$. However, these stipulations are inessential.

2.8. *Formulation of the fundamental theorem.* The theorem on the structure of finite sets of lattice points with small double sets ($T < Ck$) given below is the basic result of this book. It is essentially the first substantial result demonstrating the fruitfulness of the general ideas from Chapter I which we have illustrated there by elementary examples.

THEOREM. *Let* $K \subset Z_m, 0 \in K, T < Ck, C \geqslant 2$, *where* C *is a constant not depending on* k. *Then there exists a positive integer* n, *a homomorphism* $\varphi \colon Z_n \to Z_m$, *a canonical parallelepiped* $H \subset E_n$ *and positive constants* k_0 *and* c, *depending only on* C, *such that the following assertions are true for all* $k > k_0$:
  1) $K \subset (H \cap Z_n)\varphi$.
  2) $H \cap Z_n$ *and* $(H \cap Z_n)\varphi$ *are isomorphic.*
  3) $T(H \cap Z_n) < ck$,
  4) $n \leqslant [C - 1]$.

2.9. *A special case illustrating Theorem* 2.8. Let $K \subset Z_1$. First we consider the case $2 \leqslant C < 3$; then we have $n = 1$ by Theorem 2.8, 4); H is an interval, and by 3) the number of integers in it does not exceed ck. In view of 1) the set K is contained in an arithmetic progression whose number of elements does not exceed ck. In §1.9 we formulated a sharper theorem obtained by elementary methods.

Let $3 \leqslant C < 4$; then $n \leqslant 2$. If $n = 1$ then the set K is contained in an arithmetic progression such that the number of its elements does not exceed ck. If $n = 2$ then H is a parallelogram. Applying (1.15.4) it is easy to show that H may be taken to be a rectangle.

Thus in this case the set K is contained in a system of arithmetic progressions with identical differences whose first terms belong to an arithmetic progression with another difference.

Indeed, the images of the lattice points in a rectangle lying on different straight lines parallel to the horizontal axis form arithmetic progressions with identical differences. The image of points of the rectangle lying on some straight line parallel to the vertical axis forms a system of first points in these progressions.

2.10. *Another version of the fundamental theorem.* The formulation of Theorem 2.8 is given in a form suitable for the proof. In it we have chosen a special form of the domain containing a set isomorphic to K (a canonical parallelepiped) and we have referred to a method of obtaining an isomorphic image of K (an isomorphic mapping induced by some homomorphism of the additive group $Z_n \subset E_n$ of integral vectors into the group Z_m). These specifications are not of a basic nature. Therefore the fundamental theorem may be reworded in the following way:

THEOREM. *If $T < Ck$, $C \geqslant 2$, and $K \subset Z_m$, then there exist positive constants c and k_0, depending only on C, and a positive integer $n \leqslant [C - 1]$ such that for $k > k_0$ the set K is a subset of some set K_0 of lattice points which is isomorphic to the set of interior lattice points of a certain convex set $D \subset E_n$ and for which $T(K_0) < ck$.*

It is not difficult to derive from Theorem 2.10 the precise version of the fundamental theorem 2.8 (see § 2.29).

§4. Proof of the fundamental theorem

2.11. *A probabilistic estimate.*

LEMMA. *Let γ be a fixed real number, $\frac{1}{2} < \gamma < 1$, and let r (not necessarily independent) random variables ξ_i, $1 \leqslant i \leqslant r$, be given such that each of them assumes the two values 0 and 1. If*

$$P(\xi_i = 0) = \sum_{\substack{u_j = 0,1 \\ j \neq i \\ u_i = 0}} P_{u_1 u_2 \ldots u_r} \geqslant \gamma, \qquad 1 \leqslant i \leqslant r, \tag{2.11.1}$$

where

$$P_{u_1 u_2 \ldots u_r} = P(\xi_1 = u_1, \xi_2 = u_2, \ldots, \xi_r = u_r),$$

then for each $r > r_0(\gamma)$ there exists an r-tuple of values $u_1, \ldots, u_r$ such that

$$P_{u_1 u_2 \ldots u_r} > c(\gamma)(\sqrt{r})^{-1} c_1^r,$$

with

$$c_1(\gamma) = (1 - \gamma)^{1-\gamma} \gamma^\gamma.$$

REMARK. Since the function $(\log c_1(u))' = \log(u/(1 - u))$ increases monotonically on the interval $(0,1)$ and vanishes for $u = \frac{1}{2}$, one has

$$\min_{0 < u < 1} c_1(u) = c_1\left(\frac{1}{2}\right) = \frac{1}{2}$$

and, furthermore, $c_1(\gamma) > 1/2$.

PROOF. We introduce the notation

$$m = \max_{u_j, 1 \leqslant j \leqslant r} P_{u_1 u_2 \ldots u_r,}, \qquad s_0 = [(1 - \gamma)r + 2].$$

From (1) we obtain

$$\gamma r \leqslant \sum_{i=1}^{r} P(\xi_i = 0) = \sum_{s=1}^{r} (r + 1 - s) \sum_{\substack{u_1, u_2, \ldots, u_r \\ \sum_{i=1}^{r} u_i = s-1}} P_{u_1 u_2 \ldots u_r}$$

$$\leqslant \sum_{s=1}^{s_0} (r + 1 - s) \sum_{\sum u_j = s-1} P_{u_1 u_2 \ldots u_r} \qquad (2.11.2)$$

$$+(r - s_0) \sum_{s=s_0+1}^{r} \sum_{\sum u_i = s-1} P_{u_1 u_2 \ldots u_r} \leqslant mr \sum_{s=0}^{s_0-1} C_r^s + r - s_0.$$

For $r > r_1(\gamma)$ we have

$$\gamma r - (r - s_0) \geqslant 1,$$

and hence

$$\sum_{s=0}^{s_0-1} C_r^s < C_r^{s_0-1} \sum_{j=0}^{\infty} \left(\frac{s_0 - 1}{r - s_0 + 2} \right)^j$$

$$= C_r^{s_0-1} \frac{r - s_0 + 2}{r - s_0 + 3} < \left(\frac{\gamma}{2\gamma - 1} + 1 \right) C_r^{s_0-1}.$$

Thus (2) implies

$$m > \frac{2\gamma - 1}{3\gamma - 1} \frac{1}{r C_r^{s_0-1}}. \qquad (2.11.3)$$

By the local De Moivre-Laplace Theorem (see, for example, [1], p. 75),

$$\lim_{r \to \infty} \left(\frac{r!}{(s_0 - 1)! (r - s_0 + 1)!} (1 - \gamma)^{s_0-1} \gamma^{r-s_0+1} : \frac{e^{-(1/2)(\gamma x_1^2 + (1-\gamma)x_2^2)}}{\sqrt{2\pi\gamma(1 - \gamma)r}} \right) = 1,$$

where

$$x_1 = -x_2 = \frac{s_0 - 1 - r(1 - \gamma)}{\sqrt{\gamma(1 - \gamma)r}}.$$

Since $\lim_{r \to \infty} x_1 = 0$, there exists a constant $r_2(\gamma)$ such that for all $r > r_2(\gamma)$ the inequality

$$C_r^{s_0-1} < \frac{1}{(1 - \gamma)^{3/2} r^{1/2}} c_1^{-r} \qquad (2.11.4)$$

holds. For $r > \max(r_1, r_2)$ we obtain from (3) and (4) the relation

$$m > \frac{(2\gamma - 1)(1 - \gamma)^{3/2}}{2} r^{-1/2} c_1^r.$$

Thus the assertion of the lemma is true with

$$c(\gamma) = \frac{(2\gamma - 1)(1 - \gamma)^{3/2}}{2}, \qquad r_0(\gamma) = \max(r_1(\gamma), r_2(\gamma)).$$

2.12. *On the structure of $K \subset E_n$ for $T < Ck$, $C < 2^n$.*

LEMMA. *Let $K = \{\bar{a}_0, \ldots, \bar{a}_{k-1}\}$ be a given subset of E_n with*

$$T < Ck, 2 \leqslant C < 2^r, \qquad 1 < r \leqslant n.$$

Furthermore, let

$$\varepsilon_1 = \frac{2^r - C}{3^r (4C)^{2r}}, \qquad i_0 = \left[\frac{C-1}{\varepsilon_1}\right] + 1, \qquad k_0 = \varepsilon_1^{-i}.$$

Then for each $k > k_0$ there exists a hyperplane L with

$$T(L \cap K) \geqslant \varepsilon_1 k;$$

furthermore, L is parallel to the linear subspace $L_1(\bar{e}_1, \ldots, \bar{e}_{n-r})$ if $r < n$.

PROOF. Suppose the set K has the following property: If

$$K' \subset K, \qquad T(2K') < Dk', \qquad D \leqslant C, \qquad k' = T(K') > \varepsilon_1^{-1},$$

then there exists a set $K'' \subset K'$ for which

$$k'' = T(K'') \geqslant ck', \qquad T(2K'') < (D - \varepsilon_1)k''.$$

In this case we can construct a sequence of sets $R_0 = K$, $R_1, \ldots, R_p$ such that each R_i satisfies the conditions

$$T(2R_i) < (C - i\varepsilon_1)r_i, \qquad r_i = T(R_i), r_i > \varepsilon_1^{i - i_0}$$

and for $p = i_0$ this leads to the inequality $T(2R_p) < r_p$, which is impossible.

Thus in order to prove the lemma indirectly it suffices to show that the stated property follows from the assumption that for any hyperplane L (parallel to L_1 if $r < n$) the inequality

$$T(L \cap K) \leqslant \varepsilon_1 k$$

is satisfied.

The set $2K'$ contains $k'(k' + 1)/2$ sums of the form $\bar{a}_i + \bar{a}_j, 0 \leqslant i \leqslant j \leqslant k' - 1$. There exist more than $(k' + 1)/2D$ such sums equal to a given vector $\bar{b}_1$. Through the point $\bar{b}_1/2$ we construct an arbitrary hyperplane B_1 (parallel to L_1 if $r < n$). From the set of points which are pairwise symmetric relative to $\bar{b}_1/2$ we choose the points which do not belong to B_1 and which are located on any one side of B_1. Denoting the set so obtained by K_1, we have

$$k_1 = T(K_1) > (1/2D - \varepsilon_1)k' \geqslant k'/4D,$$

$$T(2K_1) < T < Dk' < D_1k_1; \quad D_1 = 4D^2.$$

$$(2.12.1)$$

Suppose the sets $K_1, \ldots, K_i$ have already been constructed with formula (1) included among the formulas (2). Let us assume that $K' = K_0, k' = k_0, D = D_0$, and

$$k_s = T(K_s) > k_{s-1}/2D_{s-1} - \varepsilon_1 k' \geqslant k_{s-1}/4D_{s-1},$$

$$T(2K_s) \leqslant D_s k_s, D_s = 4^{2^s-1}D^{2^s}, \qquad s = 1, 2, \ldots, i, i < r.$$

$$(2.12.2)$$

The set $2K_i$ contains $k_i(k_i + 1)/2$ sums of the form $\bar{a}_{j_r} + \bar{a}_{j_s}, 0 \leqslant r, s \leqslant k_i - 1$, $j_r \leqslant j_s$. There exist at least $(k_i + 1)/2C_i$ such sums equal to a given vector $\bar{b}_{i+1}$. Through the points $\bar{b}_{1/2}, \ldots, \bar{b}_{(i+1)/2}$ we construct an arbitrary hyperplane B_{i+1} (parallel to L_1 if $r < n$). From the set of points which are pairwise symmetric relative to $\bar{b}_{(i+1)/2}$ we select those points which do not belong to B_{i+1} and which are located on one given side of B_{i+1}, denoting the set so obtained by K_{i+1}. The relations (2) are also valid for $s = i + 1$. We continue this construction up to $s = r$. Then in view of (2) we have the inequality

$$k_r \geqslant \frac{k'}{(4D)^{2^{r-1}}(4D)^{2^{r-2}} \ldots (4D)^2 4D} > \frac{(4C)^{2^r}}{(4D)^{2^r}} \geqslant 1, \qquad (2.12.3)$$

and hence the set K_r is nonempty. This thus guarantees that the construction is possible.

The dimension of the point set generated by $\bar{b}_1/2, \ldots, \bar{b}_r/2$ is $r - 1$. Within the set K_r we select a fixed point $\bar{e}$ of maximal distance from this hyperplane B_r. The parallelepiped with the vertices

$$\frac{\bar{b}_1}{2} + \mu_1\left(\frac{\bar{b}_2}{2} - \frac{\bar{b}_1}{2}\right) + \ldots + \mu_{r-1}\left(\frac{\bar{b}_r}{2} - \frac{\bar{b}_{r-1}}{2}\right) + \mu_r\left(\bar{e} - \frac{\bar{b}_r}{2}\right), \qquad (2.12.4)$$

$$\mu_s = \pm 1, \qquad s = 1, 2, \ldots, r,$$

shall be denoted by H.

We denote by D_{i1} and D_{i2}, $1 \leqslant i \leqslant r$, the hyperplanes parallel to L_1 running through those vertices (4) for which $\mu_i = 1$ and $\mu_i = -1$ (the hyperplane D_i, $i \leqslant r - 1$, contains the point $\bar{b}_{i+1}/2$). If the point $\bar{c}_1 \in K_r$ then let $\bar{c}_2 = \bar{c}_1$ if $\bar{c}_1$ is located on the same side of $D_{r-1,1}$ with H; otherwise let $\bar{c}_2$ be the point symmetric to $\bar{c}_1$ relative to $\bar{b}_r/2$. By means of the hyperplane $D_{r-2,1}$ and the point $\bar{b}_{r-1}/2$ we obtain the point $\bar{c}_3$, etc. up to the point $\bar{c}_r$, which is obtained by means of D_{11} and $\bar{b}_2/2$. Obviously the point $\bar{c}_r \in H$ satisfies $\bar{c}_{11} \neq \bar{d}_{11}$ if $\bar{c}_1 \neq \bar{d}_1$. Therefore

$$T(H \cap K) \geqslant k_r - \varepsilon_1 k > \tfrac{1}{2} k_r > k'/2(4D)^{2^r-1}.$$

The hyperplanes D_{i1} and D_{i2}, $i = 1, 2, \ldots, r$ divide the set K' into 3^r parts $K^{(j)}$, $j = 1, 2, \ldots, 3^r$, $K^{(1)} \subset H$, excluding the points on the hyperplane. We show that for one of the sets $K^{(j)}$, $j \geqslant 2$, the inequality

$$T(K^{(j)}) = k^{(j)} \geqslant \varepsilon_2 k', \qquad T(2K^{(j)}) < (D - \varepsilon_3)k^{(j)},$$

holds with $\varepsilon_2 = \varepsilon_3 = \varepsilon_1$. Indeed, taking into account that no more than $2\varepsilon_1 rk'$ points from the set K_1 are contained in the hyperplanes D_{i1} and D_{i2}, we would otherwise obtain

$$T(2K') \geqslant \sum_{\substack{2 \leqslant j \leqslant 3^r \\ k^{(j)} \geqslant \varepsilon_3 k'}} (D - \varepsilon_3)k^{(j)} + 2^r k^{(1)}$$

$$\geqslant Dk' + \frac{2^r - D}{2(4D)^{2^r-1}} k' - 2\varepsilon_1 rDk' - 3^r D\varepsilon_2 k' - \varepsilon_3 k' \geqslant Dk'.$$

This proves the lemma.

2.13. *Embedding a convex set into a parallepiped.*

LEMMA. *Let $D \subset E_n$ be a convex, bounded, closed set with volume $V(D)$. Then there exists a parallelepiped H whose edges $\bar{x}_i$ satisfy condition (7.1) with $D \subset H$ and $V(H) \leqslant n! \, V(D)$.*

PROOF. Let L_1 and L_2 be supporting hyperplanes[3] parallel to $L(\bar{e}_1, \ldots, \bar{e}_{n-1})$ and let l be the straight line through the points $\bar{a}_1$ and $\bar{a}_2$ such that $\bar{a}_1 \in D \cap L_1$ and $\bar{a}_2 \in U \cap L_2$; furthermore, let P be the projection of D onto L_1 parallel to l, let $V(P)$ be the volume of P, and let U be the cylinder with generating lines parallel to l, height $|\bar{a}_2 - \bar{a}_1|$ and P as its lower base. The volume of U is denoted by $V(U)$. For $n = 1$ the lemma is true.

---

[3] A hyperplane L divides the space E_n into two parts. If one of these parts does not contain D but L contains points from D, then L is called a supporting hyperplane of D.

Suppose it has been proved for all values less than a given n. Then there exists an $(n-1)$-dimensional parallelpiped H_1 satisfying condition (7.1) such that $P \subset H_1$ and $V(H_1) \leqslant (n-1)! \, V(P)$. Let W be the pyramid with basis P and vertex $\bar{a}_2$. Then $V(W) \leqslant V(D)$ and $nV(W) = V(U)$; thus $V(U) \leqslant nV(D)$. Let H be the parallelepiped with base H_1 and such generating lines as the cylinder U possesses. Since $V(U)/V(P) = V(H)/V(H_1)$, this implies $V(H) \leqslant (n-1)! \, V(U) \leqslant n! \, V(D)$.

For the case $n = 2$ see Figure 4.

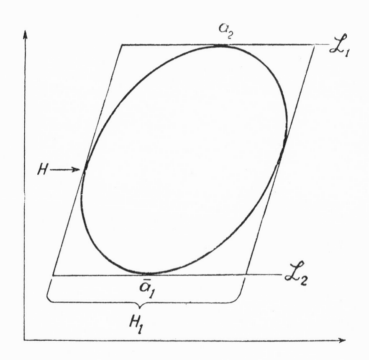

Figure 4.

REMARK. If D is a parallelepiped, we can give a precise description of the choice of the vertices of the parallelepiped H by means of the following proposition:

Let D be a given parallelepiped with linearly independent edges $\bar{x}_1, \ldots, \bar{x}_n$; furthermore, let N_i be the cylinder

$$\sigma_1 \bar{x}_1 + \sigma_2 \bar{x}_2 + \ldots + \sigma_n \bar{x}_n, \quad -\infty < \sigma_i < +\infty, \quad 0 < \sigma_j < 1, j \neq i,$$

and let F be the strip between two parallel supporting hyperplanes of D. Then an index i may be defined such that the volume $V(D_1)$ of the parallelepiped $D_1 = N_i \cap F$ satisfies the inequality $V(D_1) < nV(D)$.

Indeed, we have

$$\frac{V(D_1)}{V(D)} = \frac{\sum_{j=1}^{n} |(\bar{x}_j, \bar{e})|}{|(\bar{x}_i, \bar{e})|},$$

where $\bar{e}$ is a unit vector perpendicular to the supporting hyperplanes. This quotient does not exceed n for that value of i for which $|(\bar{x}_i, \bar{e})|$ assumes its maximal value.

2.14. *Covering a convex set by a canonical parallelepiped.*

LEMMA. *Let* $T(D \cap Z_m) = V_D$, *where D is a convex set in the space E_m. Then there exists a positive integer $n \leqslant m$, a monomorphism* $\varphi: Z_n \to Z_m$ *and a canonical parallelepiped $H \subset E_n$ such that $D \cap Z_m \subset (H \cap Z_n)\varphi$ and $V(H) < c(m)V_D$.*

PROOF. In the set U of the plane containing $D \cap Z_m$ we choose a plane L of minimal distance n (if U is empty, we let $L = E_m$). The set $D_1 = L \cap D$ is convex. Then there exists an integral edge of dimension n within the set D_1, i.e. there exist integral vectors $\bar{u}$ and $\bar{v}_1, \ldots, \bar{v}_n$, such that $\{\bar{u}, \bar{u} + \bar{v}_1, \ldots, \bar{u} + \bar{v}_n\} \subset D_1$, where $\bar{v}_1, \ldots, \bar{v}_n$ are linearly independent.

The vector $\bar{v}_1$ may be chosen in such a way that its length is minimal among those of all integral vectors on the straight line $\lambda \bar{v}_1$, $-\infty < \lambda < \infty$. Let $\bar{y}_1 = \bar{v}_1$ and $\bar{x}_i$ be integral points on $u + L(\bar{v}_1, \ldots, \bar{v}_i)$ whose distance from $\bar{u} + L(\bar{v}_1, \ldots, \bar{v}_{i-1})$ is minimal, and let $\bar{y}_i = \bar{x}_i - \bar{u}$, $2 \leqslant i \leqslant n$. The edges $\bar{y}_1, \ldots, \bar{y}_n$ thus obtained form a basis of the lattice $L \cap Z_m$.

The linear mapping $\varphi_1 : L(\bar{y}_1, \ldots, \bar{y}_n) \to E_n$ is defined by the relations $\bar{e}_i = y_i \varphi_1$, $1 \leqslant i \leqslant n$. Let $D_2 = (D_1 - \bar{u})\varphi_1$. Applying Lemma 2.13 to the set $D_2 \subset E_n$, we obtain a parallelepiped $H_1 \supset D_2$ for which $V(H_1) \leqslant n! V(D_2)$ and which satisfies condition (7.1) The method of constructing H_1 guarantees the validity of the inequalities $x_{jj} \geqslant 1$. Since a convex body which contains only one integral point must have a volume not exceeding 2^m (this follows immediately from Theorem 2.5), we have

$$V(H_1) < n! V(D_2) = n! V(D_1) < 2^m n! V_{D_1}.$$

If $L(a, b)$ is an interval of length $l = b - a \geqslant 1$, let $L_1 = (a_1, b_1)$ be a subinterval of minimal length l_1 with respect to all integers a_1 and b_1. Then it is obvious that L_1 satisfies the condition $l_1 < 2$. From this consideration it is evident that there exists a parallelepiped $H_2 \supset H_1$ with the integral vertex $\bar{v}$ and edges $\bar{x}_j$ for which the coordinates x_{jj} are integers and which satisfies the inequality $V(H_2) < 3^n V(H_1)$.

If for some j the parallelepiped H_2 thus obtained does not satisfy the condition $|x_{ij}| < x_{jj}$, then we may apply the translation φ_j defined by the following coordinate transformation:

$$y_{sj} = x_{sj} - q_{sj} x_{jj}, \qquad 1 \leqslant s \leqslant j - 1$$

$$y_{sj} = x_{sj}, \qquad\qquad j \leqslant s \leqslant n,$$

where the integers q_{sj} may be chosen in such a way that the parallelepiped $H_2 \varphi$ will satisfy the conditions already made. The parallelepiped $H = H_2 \varphi_1 \ldots \varphi_n$ is of the desired form.

2.15. *Fundamental Lemma.*

LEMMA. *Let* $K \subset Z_m$, $T(K) = k$, $T < Ck$, $C \geqslant 2$ *and* $K \subset D \subset E_m$, *where* D *is a convex set with* $T(D \cap Z_m) = V$. *Then there exist a positive integer* n, *a homomorphism* $\varphi: Z_n \to Z_m$ *and a convex set* $D_1 \subset E_n$ *such that*

$$1) \ K \subset (D_1 \cap Z_n)\varphi$$

and

$$2) \ T(D_1 \cap Z_n) < c_1 V(V/k)^{c_2}, c_1 = c_1(m, C), \tag{2.15.1}$$

$$c_2 = c_2(m, C).$$

The proof of this lemma is given in §§2.16–2.25.

2.16. *Bound for the measure of a set with large* $|S(\bar{a})|$. We consider the integral

$$W = \sum_{\bar{x}' \in K} \sum_{\bar{x}'' \in K} \sum_{\bar{x}''' \in 2K} \int_0^1 e^{2\pi i \alpha_1 (x_1' + x_1'' - x_1''')} \, d\alpha_1 \ldots \int_0^1 e^{2\pi i \alpha_m (x_m' + x_m'' - x_m''')} \, d\alpha_m$$

$$= \int_0^1 \ldots \int_0^1 S^2 S_1 \, d\alpha_1 \ldots d\alpha_m,$$

where

$$S = \sum_{\bar{x} \in K} e^{2\pi i (\bar{\alpha}, \bar{x})},$$

$$S_1 = \sum_{\bar{x} \in 2K} e^{-2\pi i (\bar{\alpha}, \bar{x})}, \quad \bar{\alpha} = (\alpha_1, \alpha_2, \ldots, \alpha_m), \quad \bar{x} = (x_1, x_2, \ldots, x_m).$$

Let I_1 be that part of the domain of integration for which

$$|S(\bar{\alpha})| > k/\sqrt{C + \varepsilon}, \qquad \varepsilon > 0. \tag{2.16.1}$$

We estimate the measure mes I_1 as follows:

$$k^2 = |W| \leqslant k^2 T(2K) \text{mes } I_1 + \frac{1}{\sqrt{C+\varepsilon}} k \int_0^1 \cdots \int_0^1 |SS_1| d\alpha_1 \ldots d\alpha_m$$

$$\leqslant CK^3 \text{mes } I_1 + \frac{\sqrt{C}}{\sqrt{C+\varepsilon}} k^2;$$

$$\text{mes } I_1 \geqslant \frac{\varepsilon}{2(C+\varepsilon)k}.$$

2.17. *Decomposition of the domain of integration in the multidimensional case.* In view of Lemma 2.14 we may assume without loss of generality that D is a canonical parallelepiped, to be denoted by H, with edges $\bar{x}_1, \ldots, \bar{x}_m$.

The domain of integration W of the integral is the cube

$$\alpha_1 \bar{e}_1 + \alpha_2 \bar{e}_2 + \ldots + \alpha_m \bar{e}_m, \qquad 0 \leqslant \alpha_i < 1, 1 \leqslant i \leqslant m.$$

We will decompose it into parts in the following manner. We introduce a decomposition of the line segment $\alpha_i \bar{e}_i$, $0 \leqslant \alpha_i < 1$, $1 \leqslant i \leqslant n$, according to the Farey series of order $Q_i = \max (1, h_i/M)$ where $h_i = x_{ii}$, $M = (V/k)^{1/4m}$. For any i each of the numbers $\alpha_i = p_i/q$, $(p_i, q_i) = 1$, $q_i \leqslant Q_i$, is contained in the Farey interval $[\alpha_i'(p_i/q_i), \alpha_i''(p_i/q_i))$. As usual, the number 0 will be considered as an element of the interval $[\alpha_i'' i(1) - 1, \alpha_i'(0))$.

Let $\bar{a} = (p_1/q_1, p_2/q_2, \ldots, p_m/q_m)$, $(p_i, q_i) = 1$, $1 \leqslant q_i \leqslant Q_i$. Through the points $\bar{a} + \alpha_i' \bar{e}_i$, $\bar{a} + \alpha_i'' \bar{e}_i$ for which all components p_i/q_i of $\bar{a}$ are less than 1 we construct a pair of hyperplanes perpendicular to $\bar{x}_i$. All m pairs of these hyperplanes are bounds for the parallelepiped $H_{\bar{a}}$. If we take as domain of integration for the integral W the union of all parallelepipeds $H_{\bar{a}}$ instead of a cube, then periodicity of the integrand implies that the integral W is not affected by this change.

We have the inequality

$$|p_i/q_i - \alpha_i'| < 1/q_i Q_i, \qquad |p_i/q_i \alpha_i''| < 1/q_i Q_i,$$

and this quantity is a lower bound for the distance between $\bar{a}$ and the planes perpendicular to $\bar{x}_i$ which we have constructed above. Figure 5 shows the two-dimensional Farey dissection, which we shall call the dissection of the cube into parallelepipeds for the case $m = 2$, $Q_1 = Q_2 = Q_3$.

2.18. *A bound for the measure of the "major intervals".* We define an interval I_1' as follows: $I_1' = \cup_{p_i,q_i} H_{\bar{a}}$, where $(p_i, q_i) = 1$, $1 \leqslant i \leqslant m$, $1 \leqslant q_i \leqslant q_0 = M^2$. Then

$$\text{mes } I_1' \leqslant \sum_{p_i,q_i} \frac{2^m}{q_1 q_2 \cdots q_m Q_1 Q_2 \cdots Q_m} \leqslant \frac{2^m}{V} M^{3m}.$$

2.19. *Reduction to rational trigonometric sums.* Suppose some vector $\bar{\alpha} \in H_{\bar{a}}$ with $\bar{\alpha} \notin I_1'$ satisfies the inequality

$$|S(\bar{\alpha})| > \frac{1}{\sqrt{C + \varepsilon}} k, \qquad \varepsilon > 0.$$

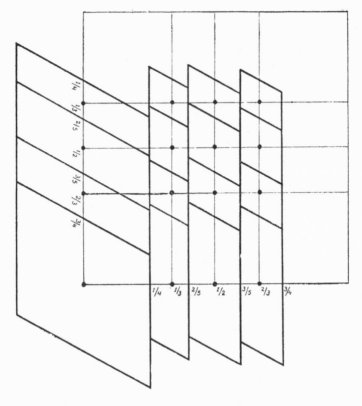

Figure 5.

Since one has

$$|(\bar{z}, \bar{x})| \leqslant \sum_{i=1}^{m} |(\bar{z}, \bar{x}_i)| < c_3 \sum_{i=1}^{m} \frac{h_i}{q_i Q_i} < \frac{c_3 m}{M},$$

this implies

$$|(S(\bar{\alpha}) - S(\bar{a}))| \leqslant \sum_{\bar{x} \in K} |e^{2\pi i(\bar{z}, \bar{x})} - 1| < c_4 \frac{k}{M}.$$

Hence, if k is sufficiently large and V is sufficiently large with respect to k, one obtains the inequality

$$|S(\bar{a})| \geqslant k/\sqrt{C + 2\varepsilon} . \tag{2.19.1}$$

2.20. *Construction of the parallelepipeds* $H_{u_1 \ldots u_r} \subset E_{m+r}$ *by means of Lemma 2.2.* In view of Lemma 2.2 the inequality (19.1) implies that for k_1 vectors $\bar{x}$ from the set K, with $k_1 > \frac{1}{2}(1 + 1/\sqrt{C + 2\varepsilon})k$, the conditions $\{(\bar{a}, \bar{x})\} = A_{\bar{x}} + \beta + \theta_{\bar{x}}, 0 \leqslant \theta_{\bar{x}} < \frac{1}{2}$, are satisfied with some integer $A_{\bar{x}}, \beta = P/Q, Q = [q_1, \ldots, q_m]$ and $0 \leqslant \beta < 1$.

We shall assume that for some positive integer r there exist r different vectors $\bar{a}_1, \ldots, \bar{a}_r$, each of which satisfy the inequality (19.1). Assuming that certain additional relations among the vectors $\bar{a}_i$ hold, a method of choosing these $\bar{a}_i, 1 \leqslant i \leqslant r$, will be described in §2.22 below.

For each i, $1 \leqslant i \leqslant r$, we may exhibit by means of (19.1) an integer $P_i, 0 \leqslant P_i < Q_i$, such that for each one of k_{1i} vectors $\bar{x}$ from the set K, where $k_{1i} > \frac{1}{2}(1 + 1/\sqrt{C + 2\varepsilon})k$, there exists an integer s, $0 \leqslant s < Q_i/2$, such that the equation

$$\frac{p_{i1}}{q_{i1}} Q_i x_1 + \frac{p_{i2}}{q_{i2}} Q_i x_2 + \ldots + \frac{p_{im}}{q_{im}} Q_i x_m + Q_i x_{m+i} = P_i + s \tag{2.20.1}$$

holds, where

$$\bar{a}_i = \left(\frac{p_{i1}}{q_{i1}}, \frac{p_{i2}}{q_{i2}}, \ldots, \frac{p_{im}}{q_{im}} \right), \qquad Q_i = [q_{i1}, q_{i2}, \ldots, q_{im}].$$

For each of the remaining $k_{2i} = k - k_{1i}$ vectors $\bar{x}$ from the set K there exists a number s, $Q_i/2 \leqslant s \leqslant Q_i$, such that (1) is satisfied.

Now we consider the hyperplanes L_{i0}, L_{i1} and $L_{i2} \subset E_{m+r}$ defined by the relation (1) for $s = 0$, $s = Q_i/2$ and $s = Q_i$, respectively. Let P_{i0} and P_{i1}, respectively, be the strip between L_{i0} and L_{i1} or L_{i1} and L_{i2} including L_{i0} or L_{i1}; furthermore, let N be the cylinder bounded by the hyperplane through each of the edges $H \subset E_m = L(\bar{e}_1, \ldots, \bar{e}_m)$ parallel to $L(\bar{e}_{m+1}, \ldots, \bar{e}_{m+r})$, and let 2^r parallelepipeds $H_{u_1 \ldots u_r}$ with $u_i = 0$ or 1, $i = 1, \ldots, r$, be defined by the equation

$$H_{u_1 u_2 \ldots u_r} = \left(\bigcap_{i=1}^{r} P_{iu_i} \right) \cap N.$$

We introduce the notation

$$H_1 = \bigcup_{\substack{u_i = 0, 1 \\ 1 \leqslant i \leqslant r}} H_{u_1 u_2 \ldots u_r} .$$

2.21. *Application of Lemmas 2.11 and 2.12.* We define a mapping $\varphi_1 : Z_{m+r} \to Z_m$

by the relation

$$(x_1, x_2, \ldots, x_{m+r})\varphi_1 = (x_1, x_2, \ldots, x_m).$$

If a point $(x_1, \ldots, x_m) \in H$ in the sense of (20.1) is given, then the point $(x_1, \ldots, x_{m+r}) \in H_1$ is uniquely determined. Therefore φ_1 is a one-to-one mapping between the integral points of these two parallelepipeds.

Applying Lemma 2.11 to the case where

$$P_{u_1 u_2 \ldots u_r} = \frac{T(K\varphi_1^{-1} \cap H_{u_1 u_2 \ldots u_r})}{k} \quad \text{and} \quad \gamma = \frac{1}{2}\left(1 + \frac{1}{\sqrt{C + 2\varepsilon}}\right)$$

we find that there exists an $H_{u_1 \ldots u_r} = H_2$ such that

$$T(K\varphi_1^{-1} \cap H_2) > c_5(\sqrt{r})^{-1} c_6^r k. \tag{2.21.1}$$

Let $\bar{x}_1, \bar{x}_2, \bar{x}_3, \bar{x}_4$ be given elements of K satisfying

$$\bar{x}_1 + \bar{x}_2 = \bar{x}_3 + \bar{x}_4, \tag{2.21.2}$$

such that there exist inverse images $\bar{y}_j = \bar{x}_j \varphi_1^{-1} \in H_{u_1 \ldots u_r}$, $1 \leqslant j \leqslant 4$. If $\bar{x}_j$ has coordinates $x_s^{(j)}$, $1 \leqslant j \leqslant m + r$, then the sth coordinate of $\bar{y}_j$ is expressible in the form

$$\frac{p_{i1}}{q_{i1}} Q_i x_1^{(j)} + \frac{p_{i2}}{q_{i2}} Q_i x_2^{(j)} + \ldots + \frac{p_{im}}{q_{im}} Q_i x_m^{(j)} + Q_i x_{m+i}^{(j)}$$
$$= P_i + s_j; \quad 1 \leqslant j \leqslant 4. \tag{2.21.3}$$

From (2) and (3) we obtain

$$Q_i(x_{m+i}^{(1)} + x_{m+i}^{(2)} - x_{m+i}^{(3)} - x_{m+i}^{(4)}) = s_1 + s_2 - s_3 - s_4.$$

Since $|s_1 + s_2 - s_3 - s_4| < Q_i$, one has $s_1 + s_2 - s_3 - s_4 = 0$ and $x_{m+i}^{(1)} + x_{m+i}^{(2)} = x_{m+i}^{(3)} + x_{m+i}^{(4)}$, $1 \leqslant i \leqslant r$; thus $\bar{y}_1 + \bar{y}_2 = \bar{y}_3 + \bar{y}_4$.

Hence we have shown that $K\varphi_1^{-1} \cap H_{u_1 \ldots u_r}$ and $(K\varphi_1^{-1} \cap H_{u_1 \ldots u_r})\varphi$ are isomorphic. Together with (1) this implies that for sufficiently large r

$$T(2(K\varphi_1^{-1} \cap H_2)) < T < Ck < 2^r k_2, \quad k_2 = T(K\varphi_1^{-1} \cap H_2).$$

Consequently, in view of Lemma 2.12, there exists a hyperplane $L \subset E_{m+r}$ such that $T(L \cap (K\varphi_1^{-1} \cap H_2)) > \varepsilon k$; $\varepsilon = \varepsilon(m, C) > 0$, where L is parallel to $L(\bar{e}_1, \ldots, \bar{e}_m)$.

2.22. *A method of choosing a system of large trigonometric sums.* We show that for any given positive integer r it is possible to choose $\bar{a}_1, \ldots, \bar{a}_r \notin I_1$ satisfying (19.1) in such a way that, for any $(m + r)$-tuple of integers A_j, $1 \leqslant j \leqslant m + r$, with

$$|A_j| \leqslant R, \quad R = (V/k)^{1/3(m+r)}, \quad \sum_{j=m+1}^{m+r} A_j^2 > 0, \qquad (2.22.1)$$

there exists at least one index i, $1 \leqslant i \leqslant m$, for which the inequality

$$|A_1 x_{1i} + A_2 x_{2i} + \ldots + A_m x_{mi} - (\bar{a}_1, \bar{x}_i) A_{m+1}$$
$$- (\bar{a}_2, \bar{x}_i) A_{m+2} - \ldots - (\bar{a}_r, \bar{x}_i) A_{m+r}| \leqslant g = (V/k)^{1/3m} \qquad (2.22.2)$$

does not hold.

Let us assume the contrary. Then there exists such an $(m + r)$-tuple of numbers A_j which satisfies the inequality (2) for all i, $1 \leqslant i \leqslant m$. In view of (1) there exists an index $j \geqslant m + 1$ such that $A_j \neq 0$ but $A_s = 0$ for $x > j$. From (2) we obtain

$$(1/|A_j|) G_{ji} - g \leqslant (\bar{a}_{j-m}, \bar{x}_i) \leqslant (1/|A_j|) G_{ji} + g, \quad 1 \leqslant i \leqslant m, \qquad (2.22.3)$$

where

$$G_{ji} = \pm (A_1 x_{1i} + A_2 x_{2i} + \ldots + A_m x_{mi} - (\bar{a}_1, \bar{x}_i) A_{m+1}$$
$$- (\bar{a}_2, \bar{x}_i) A_{m+2} - \ldots - (\bar{a}_{j-m-1}, \bar{x}_i) A_{j-1}).$$

For a fixed value of i condition (3) defines a strip of width $2g/|\bar{x}_i|$ orthogonal to $\bar{x}_i$; the system of conditions (2) for all i defines a parallelpiped H_3 whose volume does not exceed $c_7 g^m / V$. For given $\bar{a}_1, \ldots, \bar{a}_{j-m-1}$ the point $\bar{a}_{j-m}$ must be contained in this parallelepiped. Let $H(\kappa)$ be the parallelepiped which is homothetic relative to H, where the homothetic projection has its center at the center of H and a coefficient equal to κ. An arbitrary parallelepiped $H_{\bar{a}}$ with the property that $\bar{a} \in H_3$ is contained in $H_3(2)$. Indeed, the width of minimal content of $H_{\bar{a}}$ between two hyperplanes orthogonal to $\bar{x}_i$ is equal to $2/q_i Q_i \leqslant 2M/h_i < 2g/|\bar{x}_i|$, i.e. it is less than the width of the corresponding strip (orthogonal to $\bar{x}_i$) for H_3. The union P_j of the parallelepipeds $H_3(2)$ for all possible systems $A_1, \ldots, A_j$ has a volume not exceeding $c_8 g^m R^j / V$.

Now we turn to the successive construction of $\bar{a}_1, \ldots, \bar{a}_r$. Suppose the points $\bar{a}_1, \ldots, \bar{a}_{s-1}$ have already been constructed. Then we choose an $\bar{\alpha}$ satisfying condition (16.1) which is not contained in any of the parallelepipeds $H_{\bar{a}_1}, H_{\bar{a}_{s-1}}$ and which does not belong to P_s. Such a point $\bar{\alpha}$ exists since

$$V\left(\bigcup_{i=1}^{s-1} H_{\bar{a}_i} \cup P_s\right) \leqslant r\frac{M^m}{V} + c_9\frac{g^m}{V}R^{m+r} < \frac{1}{2}\ \text{mes}\ I_1.$$

If $\bar{\alpha} \in H_{\bar{a}}$ then we let $\bar{a}_s = \bar{a}$. Since $\bar{\alpha} \notin H_3(2) \subset P_s$, we have $\alpha_s \notin H_3 \subset H_3(2)$ for any $H_3(2) \subset P_s$.

2.23. *Estimation of* $T(L \cap H_1 \cap Z_{m+r})$.[4] Let L_1 be a plane of minimal dimension containing $L \cap H_1 \cap Z_{m+r}$. If the dimension of L_1 is less than $m + r - 1$ then we can find a point belonging to $H_1(2)$ such that the plane L_2 of minimal dimension containing this point and the set $L \cap H_1 \cap Z_{m+r}$ is a hyperplane. Indeed, if there exists a plane containing all points of the parallelepiped $H_1(2)$, then it must be parallel to each of the vectors $\bar{e}_1, \ldots, \bar{e}_{m+r}$. This follows from the fact that L is parallel to $L(\bar{e}_1, \ldots, \bar{e}_m)$ and that $\bar{e}_{m+1}, \ldots, \bar{e}_{m+r}$ are edges of H_1. We select an integral edge which is a basis of the lattice $L_2 \cap Z_{m+r}$, and we construct on it a parallelepiped H_4 of dimension $m + r - 1$ contained in $L_2 \cap H_1(2(n + r))$, i.e. we determine integral vectors $\bar{u}$ and $\bar{v}_1, \ldots, \bar{v}_{m+r-1}$ such that $\bar{u}, \bar{u} + \bar{v}_1, \ldots, \bar{u} + \bar{v}_{m+r-1} \in L_2 \cap H_1(2)$.

Now we consider the set $Q = \{L_3\}$ of planes parallel to L_2 which contain integral points. Let h be the minimal distance between any two of these planes. The volume of the parallelepiped generated by $\bar{v}_1, \ldots, \bar{v}_{m+r-1}$ is equal to $1/h$.

First we assume that $h < (V/k)^{-c_{10}}$, $c_{10} = 1/3(m + r)^2$. We consider a vector $\bar{w}$ which is orthogonal to $L_3 \subset Q$ and has length h. If $\bar{w} = w_1\bar{e}_1 + \ldots + w_{m+r}\bar{e}_{m+r}$, then there exists an index i such that $|w_i| \geqslant h/\sqrt{m + r}$. Then the segment of a vector parallel to $\bar{e}_i$ between two planes parallel to L_2 and at distance h from each other will have a length not exceeding $\sqrt{n + r}\, h$. Thus the number of planes from Q having a nonempty intersection with H_1 is less than c_{11}. The fundamental parallelepiped H_4 is contained in $H_5 = H_1(2(r + m))$. Therefore, if $L_4 \subset Q$ has a nonempty intersection with H_1, then the set $L_4 \cap H_5(4)$ contains a parallelepiped which is obtained from H_4 by a parallel translation; this means that it contains an integral point. Hence the set $L_4 \cap H_5(16)$ contains a figure obtained from the set $L_3 \cap H_1$ through a parallel translation by an integral vector. Consequently

$$T(L_4 \cap H_5(16) \cap Z_{m+r}) \geqslant T(L_3 \cap H_1 \cap Z_{m+r}).$$

Since

$$T(Z_{m+r} \cap H_5(16)) < c_{12}\, T(Z_{m+r} \cap H_1),$$

and furthermore

---

[4] In the paper [24] an inaccuracy occurs on p. 1040 insofar as only the case $A_j = 1$ is considered in (22.3). In the present section this argument is rectified.

$$T(Z_{m+r} \cap H_5(16)) \geqslant (c_{11}/h)T(L_3 \cap H_1 \cap Z_{m+r}),$$

we have

$$T(L_2 \cap H_1 \cap Z_{m+r}) < c_{13} Vh < c_{14} V(V/k)^{c_{10}}.$$

It remains to consider the case $h \geqslant (V/k)^{c_{10}}$. The vectors $\bar{e}_{m+1}, \ldots, \bar{e}_{m+r}$ are edges of H_1. We now refer to the remaining m edges. Since the projection of H_1 onto $L(\bar{e}_1, \ldots, \bar{e}_m)$ coincides with H, the edges of H_1 must be of the form

$$\bar{y}_j = \bar{x}_j + \alpha_{m+1j}\bar{e}_{m+1} + \alpha_{m+2j}\bar{e}_{m+2} + \ldots + \alpha_{m+rj}\bar{e}_{m+r}, \qquad 1 \leqslant j \leqslant m.$$

The vectors $\bar{y}_j$ are parallel to each vector from L_{i0}; hence they are of the form (20.1) and perpendicular to the vectors

$$\bar{u}_i = \frac{p_{i1}}{q_{i1}}\bar{e}_1 + \frac{p_{i2}}{q_{i2}}\bar{e}_2 + \ldots + \frac{p_{im}}{q_{im}}\bar{e}_m + \bar{e}_{m+i}, \qquad 1 \leqslant i \leqslant r.$$

Since

$$(\bar{y}_j, \bar{u}_i) = (\bar{x}_j, \bar{a}_i) + \alpha_{m+ij} = 0,$$

we have

$$\bar{y}_j = \bar{x}_j - (\bar{x}_j, \bar{a}_1)\bar{e}_{m+1} - (\bar{x}_j, \bar{a}_2)\bar{e}_{m+2} - \ldots - (\bar{x}_j, \bar{a}_r)\bar{e}_{m+r}, \qquad 1 \leqslant j \leqslant m.$$

We consider the lattice similar to the lattice $L_2 \cap Z_{m+r}$ with ratio $h^{1/(m+r-1)}$. The volume of a fundamental parallelepiped of this lattice F is equal to l, and the minimal distance l between the points of the lattice is not less than $h^{1/(m+r-1)}$. Let R be a sphere with its center at a point of the lattice F and radius l. Theorem 2.6 with $V \geqslant cl^{m+r-1} \geqslant ch$ and $\lambda_1 = 1$ implies that $\lambda_i \leqslant 2^{m+r-1}/V$. Thus there exists a basis in F consisting of vectors whose length does not exceed $c_{17} h^{1/(m+r-1)-1}$. This means that the set $L_2 \cap Z_{m+r}$ contains a basis $\bar{u}_1, \ldots, \bar{u}_{m+r-1}$ consisting of vectors whose length does not exceed ch^{-1}

If $f = \max V_i$. where V_i is the volume of the parallelepiped generated by $\bar{y}_i$ and $\bar{u}_1, \ldots, \bar{u}_{m+r-1}$, then there exist at least $[f]$ planes $L \subset Q$ whose intersection with H_1 is nonempty. Indeed, if h_i is the length of the projection of $\bar{y}_i$ onto the perpendicular to L_2, then $V_i = h_i/h$. Hence we have

$$V_i = \pm \begin{vmatrix} u_{11} & u_{12} & \cdots & u_{1\,m+r} \\ u_{21} & u_{22} & \cdots & u_{2\,m+r} \\ \cdots & \cdots & \cdots & \cdots \\ u_{m+r-1,\,1} & u_{m+r-1,\,2} & \cdots & u_{m+r-1,\,m+r} \\ y_{1l} & y_{2l} & \cdots & y_{m+ri} \end{vmatrix} =$$

$$= |A_1 x_{1i} + A_2 x_{2i} + \ldots + A_m x_{mi} - (\bar{a}_1, \bar{x}_i) A_{m+1}$$
$$- (\bar{a}_2, \bar{x}_i) A_{m+2} - \ldots - (\bar{a}_r, \bar{x}_i) A_{m+r}|,$$

where $|A_i| < c_{18} h^{-(m+r-1)}$ and $\sum_{j=m+1}^{m+r} A_j^2 \neq 0$, since L_2 is parallel to the vectors $\bar{e}_1, \ldots, \bar{e}_m$, i.e. $\sum_{j=1}^{m} A_j^2 = 0$. By the method of choosing the $\bar{a}_j$, $1 \leqslant j \leqslant r$, as given in §2.22 there exists an index i with $V_i \geqslant g$ such that we obtain the relation

$$T(L_2 \cap H_1 \cap Z_{m+r}) < c_{19} V g^{-1} < c_{20} V(V/k)^{-c_{21}}$$

also in the case where $h \geqslant (V/k)^{c_{10}}$.

2.24. *Estimation of the number of hyperplanes containing inverse images of vectors from K.* We shall establish an upper bound for the number of elements in the set $L_3 \subset Q$ for which $T(L_3 \cap H_1 \cap K\varphi_1^{-1}) > 0$. There exist sets $H^0_{u_1 \ldots u_r}$ and L_5 for which $T(L_5 \cap H^0 \cap K\varphi_1^{-1}) \geqslant \varepsilon_1 k/2^m$. The pair of sets $H^{(1)}_{u_1 \ldots u_r}$, $H^{(2)}_{u_1 \ldots u_r}$ is isomorphic in the sense of §1.4 to the pair $(H^{(1)} \cap Z_{m+r})\varphi_1$, $(H^{(r)} \cap Z_{m+r})\varphi_1$. Let r_1 be the number of planes $L_3 \subset Q$ whose intersection with $H_{u_1 \ldots u_r}$ contains points which correspond to points from K. Then $T \geqslant r_1 \varepsilon_1 k/2^m$. Therefore $r_1 \varepsilon_1 /2^m \leqslant C$ and hence $r_1 \leqslant 2^m C/\varepsilon_1$. In H_1 there are never more than $r_2 \leqslant 4^m C/\varepsilon_1$ planes $L^{(1)}, \ldots, L^{(r_2)} \subset Q$ whose intersection with H_1 contains points which correspond to points from the set K.

2.25. *Conclusion of the proof of the lemma.* In view of the argument developed in 2.23 and 2.24 we have

$$\left(\left(\bigcup_{s=1}^{r_2} L^{(s)}\right) \cap H_1 \cap Z_{m+r}\right)\varphi_1 \supset K,$$

and furthermore

$$T(L_s^{(s)} \cap H_1 \cap Z_{m+r}) < c_{22} V(V/k)^{-c_{23}}, \qquad 1 \leqslant s \leqslant r_2.$$

By means of Lemma 2.14 (with φ_1 as the identity mapping) each of the r_2 similar parallelepipeds $H'' \cap L_s$, $1 \leqslant s \leqslant r_2$, is included in an $(m + r - 1)$-dimensional canonical parallelepiped $H^{(s)}$ with heights $h_1^{(s)}, \ldots, h_{m+r-1}^{(s)}$. Let H_6 be a canonical parallelepiped with edges parallel to the edges of $H^{(s)}$ and $h_{i6} = \max h_i^{(s)}$, $1 \leqslant i \leqslant m + r - 1$.

Let the vertices of the parallelepipeds $H^{(s)}$ be denoted by $\bar{a}_1, \ldots, \bar{a}_s$. Then we define a homomorphic mapping $\varphi_2 : E_{m+r-1+r_1} \to E_{m+r}$ by the relations $\bar{e}_i \varphi_2 = \bar{e}_i$, $1 \leqslant i \leqslant m + r - 1$; $\bar{e}_{m+r-1+s} \varphi_2 = \bar{a}_s$, $1 \leqslant s \leqslant r_1$.

Let H_7 be a canonical parallelepiped with edges $\bar{x}_1^{(s)}, \ldots, \bar{x}_{m+r-1}^{(s)}, 2\bar{e}_{m+r}, \ldots, 2\bar{e}_{m+r-1+r_1}$, where the $\bar{x}_j^{(s)}$ are the edges of H_6. Then we have

$$(H_7 \cap Z_{m+r-1+r_1})\varphi_2 \varphi_1 \supset K, \qquad V(H_7) < c_{24} V(V/k)^{-c_{23}}.$$

It remains to let $H = H_7$, $n = m + r - 1 + r_1$ and $\varphi = \varphi_2 \varphi_1$. This completes the proof of Lemma 2.15.

2.26. *Transition to an isomorphic set.*

LEMMA. *Let* $\varphi : Z_n \to Z_m$ *be a given homomorphism with kernel N such that Z_n / N is a torsion-free group, and let $H \subset E_n$ be a canonical parallelepiped. Then there exists a homomorphism $\varphi_1 : Z_s \to Z_m$, $s \leqslant n$, and a canonical parallelepiped $H_1 \subset E_s$ with the following properties*:

1) $(H_1 \cap Z_s)\varphi_1 \supset (H \cap Z_n)\varphi$.
2) $(H_1 \cap Z_s)\varphi_1$ *and* $H_1 \cap Z_s$ *are isomorphic.*
3) $V(H_1) < c(n)V(H)$.

PROOF. If the sets $H \cap Z_n$ and $(H \cap Z_n)\varphi$ are not isomorphic (otherwise there is nothing to prove), then there exist two integral points $\bar{a}, \bar{b} \in 2H$ such that $\bar{a}\varphi = \bar{b}\bar{\varphi}$. Let $\bar{d}_1$ be an integral vector of minimal length among the integral vectors which are collinear with $\bar{b} - \bar{a}$. Since Z_n / N is a torsion-free group, one has $\bar{d}_1 \varphi = 0$. For the lattice Z_n we construct a basis containing $\bar{d}_1$ by the following method: Let i be an index with $(\bar{d}_1, \bar{e}_1) \neq 0$. In the parallelepiped with edges $\bar{e}_1, \bar{e}_2, \ldots, \bar{e}_{i-1}$, $\bar{e}_{i+1}, \ldots, \bar{e}_n, \bar{d}_1$ (or on its boundary) we choose a point $\bar{d}_2$ of minimal distance from $L_1(\bar{e}_1, \bar{e}_2, \ldots, \bar{e}_{i-1}, \bar{e}_{i+1}, \ldots, \bar{e}_{n-1}, \bar{d}_1)$ but not belonging to L_1. In the parallelepiped with edges $\bar{e}_1, \bar{e}_2, \ldots, \bar{e}_{i-1}, \bar{e}_{i+1}, \ldots, \bar{e}_{n-1}, \bar{d}_1$ we choose a point $\bar{d}_3$ with minimal distance from $L_2(\bar{e}_1, \bar{e}_2, \ldots, \bar{e}_{i-1}, \bar{e}_{i+1}, \ldots, \bar{e}_{n-2}, \bar{d}_1)$, etc.

We apply Lemma 2.14 to the parallelepiped H, letting $\bar{y}_{n+1-i} = d_i$, $1 \leqslant i \leqslant n$, in the definition of φ. We obtain a canonical parallelepiped H' for which $(H' \cap Z_n)\varphi \supset H \cap Z_n$. Let h_n be the nth height of H' and let P be the volume of its base. Then $h_n V(P) < c_1 V(H)$.

We consider the planes $x_n = t$, $t = a, \ldots, a + h_n - 1$, each of which has a nonempty intersection P_t of volume $V(P_t) = \bar{P}$ with the set H'. Let $\bar{P}_t$, $1 \leqslant t \leqslant h_n$, be the projection of P_t onto $L(\bar{e}_1, \ldots, \bar{e}_{n-1})$ and let Q be the convex hull of the set $\cup_{t=1}^{h_n} \bar{P}_t$.

The method of choosing d_1 implies the existence of an index i such that $2\bar{P}_i \cap 2\bar{P}_{i+1}$ is nonempty. Hence we have $V(Q) < c_2 h_n V(P) < c_3 V(H)$. By means of Lemma 2.13 we include Q in a parallelepiped H'' for which condition (2.7.1) is satisfied. Since $\bar{x}_{jj} \geqslant 1$ for this parallelepiped (because $Q \supset \bar{P}_a$. where P_a is an $(n - 1)$-dimensional canonical parallelepiped), H'' must be contained in a canonical parallelepiped H''' with $V(H''') < c_4 V(H'')$ (see the final part of the proof of Lemma 2.14).

Let $\bar{v} \in Z_n$, and let $\bar{\bar{v}}$ be the projection of $\bar{v}\varphi$ onto $L(\bar{e}_1, \ldots, \bar{e}_{n-1})$. The homomorphism $\varphi: Z_n \to Z_m$ induces a homomorphism $\varphi_1 : Z_{n-1} \to Z_m$ such that if $\bar{v}\varphi = \bar{v}_1 \in Z_m$, then $\bar{\bar{v}}\varphi_1 = \bar{v}_1$. If $(H''' \cap Z_{n-1})\varphi$ is not yet isomorphic with $H''' \cap Z_{n-1}$, we repeat once more the entire argument, etc.

2.27. *Sharpened version of Lemma 2.26.*

LEMMA. *Let* $\varphi: Z_n \to Z_m$ *be a given homomorphism with kernel* N *such that* Z_n/N *is a torsion-free group, and let* $H \subset E_n$ *be a canonical parallelepiped; furthermore, let* p *be a positive integer,* $p \geqslant 2$. *Then there exist a homomorphism* $\varphi_1 : Z_s \to Z_m$, $s \leqslant n$, *and a canonical parallelepiped* $H_1 \subset E_s$ *with the following properties*:
1) $(H_1 \cap Z_s)\varphi_1 \supset (H_1 \cap Z_n)\varphi$.
2) *There does not exist any pair* $\bar{a}$, $\bar{b}$ *of integral points such that*

$$\bar{a}, \bar{b} \in pH, \qquad \bar{a}\varphi_1 = \bar{b}\varphi_1.$$

3) $V(H_1) < c(n,p)V(H)$.

In §2.26 this lemma was proved for the case $p = 2$. The proof for arbitrary $p \geqslant 2$ does not require any change.

2.28. *Proof of Theorem* 2.8. We construct a sequence of homomorphisms φ_i: $Z_{n_i} \to Z_m$ and canonical parallelepipeds $H_i \subset E_{n_i}$ such that φ_i, H_i and n_i satisfy conditions 1), 2) and 4) of the theorem for any i and also the inequality $V_i < V_{i-1}/2$ for $i \geqslant 2$.

For $i = 1$ we choose as φ_i the identical mapping $\varphi_1 : Z_m \to Z_m$, and as $H_1 \subset E_n$ we choose, for example, one of the canonical parallelepipeds of minimal volume among those containing the sets K. Suppose this sequence has already been defined for all i with $1 \leqslant i \leqslant s$. Applying Lemmas 2.15 and 2.26 we obtain a homomorphism $\varphi_{s+1} : Z_{n_{s+1}} \to Z_{n_s}$ and a canonical parallelepiped $H_{s+1} \subset E_{n_{s+1}}$ such that the assumptions 1) and 2) of the theorem are satisfied.

We may assume that the set of inverse images of the points of K contained in the set H_{s+1} be of dimension n_{s+1}, since otherwise we could lower its dimension by applying Lemma 2.14 to the set $H_{s+1} \cap L$ (where L is a hyperplane containing the inverse images of the points from K). Therefore condition 4) is satisfied in view of Lemma 1.14.

By condition 2) of Lemma 2.15 we have $V_{s+1} < c_1 V_s(V_s/k)^{-c_2}$, where c_1 and c_2 are positive constants depending only on C. We can choose a sufficiently large positive constant c_3 such that the inequality $V_{s+1} < V_s/2$ is satisfied for all $V_s > c_3 k$. We continue the construction of the sequence up to an index i_0 for which $V_{i_0} < c_3 k$. Then $\varphi = \varphi_{i_0}$ and $H = H_{i_0}$ have the desired properties, since the last condition 3) is also satisfied.

REMARK. For sets K with $a_0 = 0$ and $a_{k-1} < ck$, where c is the constant in condition 3) of Theorem 2.8, this theorem obviously does not furnish any additional information on the structure of these sets. Here the elementary method yields sharper results for small values of T ($T \leqslant 3k - 3$).

Thus the method of trigonometric sums is applicable to the investigation of sufficiently "thin" sets, just as for direct additive problems the method of trigonometric sums furnishes results only for sufficiently large numbers.

2.29. *Relation between the different versions of the fundamental theorem.* If we assume that the set K has dimension m and $K\varphi^{-1} \cap H$ has dimension n (which is the case, in view of the argument given in §2.28), then the inequality

$$m \leqslant n \tag{2.29.1}$$

follows from the fact that a homomorphic mapping preserves linear dependence.

Applying the fundamental theorem to an isomorphic mapping of K into E_r, and using the inequality (1), we obtain Theorem 1.23.

Conversely, Theorem 2.8 may be derived from Theorem 1.23 by applying Lemma 2.14 and Theorem 1.26. Thus Theorem 2.8 follows from Theorem 2.10.

§5. Additive properties of nonaveraging sets

2.30. *Additive properties of nonaveraging sets.* K. F. Roth [38] showed that if a set K does not contain any arithmetical progression of length 3, i.e. if

$$a_i + a_t \neq 2a_j, \tag{2.30.1}$$

for any $a_i, a_j, a_t \in K$, then

$$\lim_{k \to \infty} \frac{a_{k-1} - a_0}{k} = +\infty. \tag{2.30.2}$$

For such nonaveraging sets the following theorem is true.

THEOREM. *Let* $K \supset \{0\}$ *be a nonaveraging set. Then*

$$\lim_{k \to \infty} \frac{T}{k} = +\infty.$$

PROOF. Suppose there exists a constant $C > 0$ such that $T < Ck$. We consider the canonical parallelepiped $H \subset E_n$ obtained by applying Theorem 2.8. Then we include H in a parallelepiped H_1 with edges parallel to the edges of H, with the origin as its center of symmetry and with volume $V(H_1) \leqslant 2^n V(H)$. Since $H_2 = c_1 k^{-1/n} H_1$ has a volume larger than 2^n if c_1 is sufficiently large, there exists in H_2 a lattice point different from the origin, and H_1 contains some segment of a straight line l on which at least $2k^{1/n}/c_1$ lattice points are located.

We construct a basis $\bar{a}_1, \ldots, \bar{a}_n$ of the lattice Z_n such that $\bar{a}_n$ and l are parallel. Let H_3 be a parallelepiped which is canonical with respect to this basis, containing

H_1 and with the property that $V(H_3) \leqslant n! \, V(H_1)$ (see the Remark following Lemma 2.13). Then the base of H_3 is a parallelepiped H_4 which is canonical relative to the basis $\bar{a}_1, \ldots, \bar{a}_{n-1}$ and has a volume not exceeding $c_2 k^{1-1/n}$. Thus we have found that all lattice points of the parallelepiped H are contained in line segments whose number is $V(H_4)$, and the number of lattice points in each of these segments is $V(H_3)/V(H_4)$. Since $k \leqslant V(H_3) < c_4 k$ and $V(H_4) \leqslant c_2 k^{1-1/n}$, we have the inequality

$$V(H_3)/V(H_4) \geqslant c_3 k^{1/n}. \tag{2.30.3}$$

Within H there are k inverse images of the set K. Therefore there exists a segment which contains at least $k/V(H_4) > V(H_3)/c_4 V(H_4)$ inverse images. By (3) and (2) there exist three points $\bar{b}_1, \bar{b}_2, \bar{b}_3$ which are inverse images of points from K satisfying $\bar{b}_1 + \bar{b}_3 = 2\bar{b}_2$. This contradicts condition (1) of the theorem.

CHAPTER III
SUMS OF SEQUENCES, SETS OF RESIDUES AND POINT SETS

The present chapter is devoted to applications. We will obtain a sharpening of a theorem of Kneser for sums of sequences with positive asymptotic density and we shall prove a conjecture of Erdös for sequences with density zero.

Furthermore we shall investigate the structure of sets of residues modulo a prime number and of point sets of positive measure in n-dimensional euclidean space.

§1. Brief summary of known results

3.1. *Notation.* Let A be an increasing sequence of nonnegative integers:

$$A = \{a_0, a_1, \ldots, a_i, \ldots\}, \, a_{i+1} > a_i, \, i \geqslant 0,$$

$$d(A) = \lim_{k \to \infty} (a_1 - a_0, a_2 - a_0, \ldots, a_k - a_0),$$

where $(a_1 - a_0, a_2 - a_0, \ldots, a_k - a_0)$ denotes the greatest common divisor of the numbers $a_1 - a_0, a_2 - a_0, \ldots, a_k - a_0$.

The number of positive elements in the sequences A and $2A$ not exceeding x will be denoted, respectively, by $A(x)$ and $A_2(x)$.

By $\alpha' = \alpha'(A)$ we denote the Šnirel'man density of the sequence A, defined by the equation

$$\alpha' = \inf_{x \in N} \frac{A(x)}{x},$$

where N is the set of positive integers. By $\alpha = \alpha(A)$ we denote the asymptotic density of the sequence A, defined by the equation

$$\alpha = \lim_{x \to \infty} \frac{A(x)}{x};$$

furthermore we use the notation $\gamma = \alpha(2A)$.

3.2. *Metric additive number theory.* L. G. Šnirel'man [41] initiated the study of additive properties for the very limited class of sequences of integers with positive Šnirel'man density. For sums of two identical sequences A. Ja. Hinčin [26] obtained the following results:

If $\alpha'(A) = \alpha$, $\alpha'(2A) = \gamma$ and $a_0 = 0$ then

$$\gamma \geqslant \min(2\alpha, 1). \tag{3.2.1}$$

H. B. Mann [32] gave a proof for an analagous theorem in the case where two different sequences are added.[1]

M. Kneser [29] obtained an analog of Mann's result for sums of sequences with positive asymptotic density.

We give a somewhat modified version of Kneser's theorem for the sum-sets of two identical sequences:

THEOREM. *Either the inequality* $\gamma \geqslant \min(2\alpha, 1)$ *holds or there exist positive integers* g *and* h, *and residues* $r_1, \ldots, r_h \bmod g$ *such that*

$$T(2\{r_1, r_2, \ldots, r_h\}) = 2h - 1,$$

$$\alpha > (2h - 1)/2g, \tag{3.2.2}$$

and A is contained in the union of the h residue classes determined by the residues $r_1, \ldots, r_h$.

In [13] Erdös formulated Conjecture 15 as an analog to Hinčin's theorem for sequences of density zero. He conjectured that the inequality

$$\delta = \varlimsup_{x \to \infty} \frac{A_2(x)}{A(x)} \geqslant 3$$

holds for any set A satisfying

$$\lim_{x \to \infty} \frac{A(x)}{x} = 0.$$

Similar theorems have also been proved for sums of sets of residues modulo a prime (Cauchy [9]; Davenport [10], §2.1), and for sums of point sets with positive measure (Henstock and Macbeath [25]).

In the present chapter we obtain sharpened versions of the majority of the results mentioned.

3.3. *Work on inverse additive problems.* The concept of an inverse additive problem (§1.7) enables us to consider from a single point of view many published papers devoted to the solution of various specific questions in additive number theory.

The paper [12] by Erdös and our papers [14] and [15] are devoted to inverse additive problems of the following type for the representation of integers by an unbounded number of terms:

Let $q(u)$ be the number of solutions of the inequality

$$a_1 n_1 + a_2 n_2 + \ldots + a_r n_r + \ldots \leqslant u$$

---

[1] A formulation and a simple proof of Mann's theorem may be found in [27].

in terms of nonnegative integers n_i, where the a_i are elements of a given set A. Let $q(u)$ be given. What can be said about $A(u)$?

In [12] Erdös also considered the following inverse problem: Let $q(u)$ be the number of solutions of the inequality

$$a_i + a_j \leqslant u,$$

and suppose that $\lim_{u \to \infty} q(u)/cu^{2\alpha} = 1$. Then

$$\lim_{u \to \infty} A(u)/c_1 u^\alpha = 1, \qquad c_1 = c_1(c).$$

In 1960 V. Tašbaev [42] investigated the question of the remainder term in this inverse problem.

Let us turn to the relation between inverse problems and questions on the distribution of primes. If we let $q(u) = [e^u]$, then $a_i = \ln p_i$. where p_i is the ith prime number. Therefore the problem of the distribution of primes may be considered as an inverse problem of additive number theory of the form given (Beurling [2]; Bredihin [3]–[6]). It is easy to reformulate Kneser's theorem in such a form that it gives a solution to an inverse problem: If $\alpha(A) = \alpha > 0$ and $\alpha(2A) < 2\alpha$, $\alpha \leqslant \frac{1}{2}$, then the set A has the structure described in the preceding theorem.

Whereas Kneser's theorem appears to be of final character in its first formulation, the new formulation gives rise to the natural question on sharpening it as follows: If $\alpha(A) = \alpha$ and $\alpha(2A) < C\alpha$, $C \geqslant 2$, what can be said about the structure of A? (See §3.6.)

A solution of inverse problems in additive number theory is obtained from investigations which clarify in which cases lower bounds in inequalities of direct metrical problems are actually attained.

In 1956 A. G. Vosper [43] investigated sums of subsets of the additive group of residue classes modulo a prime. Theorem 2.1 is a sharpening of his results.

H. B. Kemperman [28] solved an inverse problem for abelian groups. He determined the structure of subsets B and C of an abelian group with b and c elements, respectively, under the condition that the number of elements in the set $B + C$ does not exceed $b + c - 1$. Kemperman's work is a natural extension of the work of Kneser cited above which involves the additive group of residues of an arbitrary modulus, whereas Vosper solved the corresponding problem for the additive group of residues modulo a prime. In the work of Henstock and Macbeath [25] the form of point sets with positive measure is determined under the condition that the measure of their sum-sets assumes the smallest possible value (given by the Brunn-Minkowski inequality).

The basic part of the results contained in the paper [16] through [24] is the subject of the present book.

§2. Sums of sequences with positive asymptotic density

3.4. *An elementary method.* P. Erdös [11] has shown that the following result is implied by Hinčin's theorem: If $0, 1 \in A$ then $\gamma \geqslant \min(1, 3\alpha/2)$. We will now show that a necessary and sufficient condition for the inequality $\gamma \geqslant 3\alpha/2$ may be obtained from the elementary Theorem 1.9 in the form $\alpha \leqslant 2/3d(A)$.

THEOREM.

$$\gamma \geqslant \min(1/d(A), 3\alpha/2).$$

PROOF. Without loss of generality we assume that $a_0 = 0$ and $d(A) = 1$.

If $\alpha > \frac{1}{2}$, then the set $2A$ contains all integers from a certain value onwards, and thus $\gamma = 1$. Therefore we assume that $\alpha \geqslant \frac{1}{2}$ and $d(A) = 1$. In order to prove the assertion indirectly let us assume that $\gamma < 3\alpha/2$.

We define a sequence of positive integers $y_1 < y_2 < \ldots$ with the property

$$\lim_{j \to \infty} \frac{A_2(y_j)}{y_j} = \gamma.$$

Then there exists for any positive ε an index $j_0(\varepsilon)$ such that the inequalities

$$A_2(y_j) < (\gamma + \varepsilon)y_j, \tag{3.4.1}$$

$$A(y_j/2) > (\alpha - \varepsilon)y_j/2 \tag{3.4.2}$$

are satisfied for all $j > j_0(\varepsilon)$.

Let M be the smallest number i such that $a_i > y_j/2$. Then $a_{M-1} \leqslant y_j/2$. We use the notation

$$\Delta = a_M - y_j/2, \qquad \Delta_1 = y_j/2 - a_{M-1}.$$

If we assume that $a_M \leqslant 2M - 1$, then Theorem 1.3 implies

$$A_2(y_j) \geqslant a_M + M - \Delta - 1 = (a_M - \Delta) + (M - 1),$$

and therefore by (1) and (2) we have

$$(\gamma + \varepsilon)y_j > y_j/2 + (\alpha - \varepsilon)y_j/2$$

and finally

$$\gamma \geqslant (1 + \alpha)/2 \geqslant 3\alpha/2.$$

Thus we may assume that

$$a_M > 2M - 1. \tag{3.4.3}$$

For sufficiently small values of ε the inequalities (1) and (2) imply

$$\frac{A_2(y_j)}{A(y_j/2)} < 2\frac{\gamma + \varepsilon}{\alpha - \varepsilon} < 3.$$

Thus by Theorem 1.9

$$a_{M-1} < 2M - 3.$$

Again applying Theorem 1.9, we obtain

$$(\gamma + \varepsilon)y_j > A_2(y_j) \geqslant a_{M-1} + M \geqslant a_{M-1} + (\alpha - \varepsilon)a_M$$
$$= y_j/2 - \Delta_1 + (\alpha - \varepsilon)(y_j/2 + \Delta). \tag{3.4.4}$$

Therefore, since $\alpha \leqslant \frac{1}{2}$ and $\alpha < 3\gamma/2$, one has

$$\Delta < 2\Delta_1 \tag{3.4.5}$$

for all sufficiently small values of ε. By the inequality (3) and Theorem 1.9,

$$(\gamma + \varepsilon)y_j \geqslant A_2(y_i) \geqslant 3M - (\Delta - \Delta_1) - 3 \geqslant 3(\alpha - \varepsilon)a_M - (\Delta - \Delta_1)$$
$$= 3(\alpha - \varepsilon)y_j/2 - (\Delta - 3(\alpha - \varepsilon)\Delta - \Delta_1) \tag{3.4.6}$$

If $(1 - 3\alpha)\Delta < \Delta_1$ (for $\alpha > 1/6$ this follows from (5)), then (6) implies $\gamma \geqslant 3\alpha/2$. If $(1 - 3\alpha)\Delta \geqslant \Delta_1$ and therefore $\alpha \leqslant 1/6$, then by (4) we have $\Delta/y_j > 1/2$, and hence $(\gamma + \varepsilon)y_j > A_2(y_j) \geqslant 2A(y_j/2) - 1 = 2A(y_j) - 1 \geqslant 2(\alpha - \varepsilon)y_j - 1$. Consequently $\gamma \geqslant 2\alpha$.

3.5. *An auxiliary lemma from the geometry of numbers.*

LEMMA. *Let $\Gamma \subset E_n$ be a given lattice whose fundamental parallelepiped has volume one; furthermore let R be a closed convex set which is symmetric with respect to the origin O and which does not contain any two points which are congruent mod Γ. Then there exists a basis $\bar{a}_1, \ldots, \bar{a}_n$ of the lattice Γ such that $R \subset H$, where H is the parallelepiped with edges $c\bar{a}_1, \ldots, c\bar{a}_n$, $c = 2n!$, and with its center at O.*

PROOF. Let $\bar{a}_1$ be an arbitrary lattice point different from the origin such that $a_1 \in \lambda_1 R$ (for the definition of the numbers λ_i see 2.6). If $\bar{a}_1, \ldots, \bar{a}_{i-1}$ have been defined, we choose as $\bar{a}_i$ any lattice point within the set $\lambda_i R$ which is not a linear combination of the points $\bar{a}_1, \ldots, \bar{a}_{i-1}$ (with integer coefficients). The vectors $\bar{a}_1, \ldots, \bar{a}_n$ form a basis of the lattice Γ. Let the linear transformation $\varphi: E_n \to E_n$ be defined by the relations $\bar{a}_i\varphi = \bar{e}_i$, $1 \leqslant i \leqslant n$. The set $R_1 = R\varphi$ is then subjected to a dilatation in the direction $\bar{e}_1$ in such a way that $\bar{e}_1$ becomes a boundary point, a dilatation in the direction $\bar{e}_2$ in such a way that $\bar{e}_2$ becomes a boundary point, etc. Finally we obtain a set R_2 whose volume, by Theorem 2.5, does not exceed 2^n.

Furthermore,

$$V(R_2 \cap L(\bar{e}_1, \bar{e}_2, \ldots, \bar{e}_{i-1}, \bar{e}_{i+1}, \ldots, \bar{e}_n)) \geqslant 2^{n-1}/(n-1)!, \quad 1 \leqslant i \leqslant n,$$

since the convex hull of the points O, $\bar{e}_1, \ldots, \bar{e}_{n-1}$ has volume $1/(n-1)!$. Therefore, if R_1 contains a point whose distance from $L(\bar{e}_1, \ldots, \bar{e}_{i-1}, \bar{e}_{i+1}, \ldots, \bar{e}_n)$ is h, then

$$V(R) \geqslant 2h \cdot 1/n \cdot 2^{n-1}/(n-1)! = (2^n/n!)h.$$

But since R does not contain any two points which are congruent mod Γ, this implies $V(R) = V(R_1) \leqslant V(R_2) \leqslant 2^n$. Therefore $h \leqslant n!$.

3.6. *Sums of sets of positive density.*

THEOREM.[2] *Let A be a set with $a_0 = 0$ and $d(A) = 1$ which satisfies the conditions $\alpha > 0$ and $\gamma < C\alpha$, $C \geqslant 3/2$; furthermore assume that there exists a constant C_i such that there exist only finitely many indices i for which the inequality*

$$a_{i+1} > C_1 a_i \tag{3.6.1}$$

holds. Then there exist a cylinder $N \subset E_n$, $n \geqslant 2$, whose base is a convex set $P \subset L(\bar{e}_1, \ldots, \bar{e}_{n-1})$, and positive constants c and α_0, depending only on C and C_1, such that for $\alpha < \alpha_0$ the following assertions are true:
1) $A \subset (N \cap Z_n)\varphi$,
2) $N \cap Z_n$ *and* $(N \cap Z_n)\varphi$ *are isomorphic,*
3) $V(P) < c\alpha$, $T(N^{(x)} \cap Z_n) < c\alpha x$,
4) $n \leqslant [2\,C - 1]$.
Here $N^{(x)}$ is the subset of N consisting of the points for which $x_n \leqslant x$; the mapping $\varphi: Z_n \to Z_1$ is a homomorphism such that $\bar{e}_i\varphi = 0$, $1 \leqslant i \leqslant n - 1$, and $\bar{e}_n\varphi = 1$.

PROOF. First we remark that, since $d(A) = 1$ and α_0 is sufficiently small, Theorem 3.4 implies the inequality $\gamma \geqslant 3\alpha/2$, which accounts for the stipulation imposed on C in the conditions of the theorem.

By the inequality

$$\gamma = \lim_{x \to \infty} \frac{A_2(x)}{x} < C\alpha$$

there exists, for any given $\varepsilon > 0$, a sequence of positive integers

---

[2] Here we correct an error which occurs in the wording of this theorem in [23]. No change in the proof is needed.

$$y_1 < y_2 < \cdots < y_j < \cdots, \tag{3.6.2}$$

such that, as in (4.1), the inequality

$$A_2(y_j)/y_j < \gamma + \varepsilon \tag{3.6.3}$$

holds. We let $K_j = \{a_i \leqslant y_j/2\}$.

By the definition of the number α we may assume that, as in (4.2),

$$A(y_j/2) = T(K_j) > (\alpha - \varepsilon)y_j/2. \tag{3.6.4}$$

Furthermore, it is evident that

$$T(2K_j) \leqslant A_2(y_j). \tag{3.6.5}$$

By (3), (4) and (5) we may choose a sufficiently small number ε such that

$$\frac{T(2K_j)}{T(K_j)} \leqslant 2\frac{A_2(y_j)}{y_j} : \frac{T(K_j)}{y_j/2} < \frac{2(\gamma + \varepsilon)}{\alpha - \varepsilon} < 2C. \tag{3.6.6}$$

In view of (6) we may apply Theorem 2.8 to each of the sets K_j in the case where $T(K_j) > k_0$, where k_0 is defined by the quantity $2C$, as stated in the conditions of Theorem 2.8. In order that this condition be satisfied for each j we only have to choose y_1 sufficiently large, which is obviously always possible.

As a result of applying Theorem 2.8 to each j there exists a positive integer n_j, a homomorphism $\varphi_j : Z_{n_j} \to Z_1$ and a canonical parallelepiped H_j satisfying conditions 1)–4) of Theorem 2.8. We may assume that all n_j are equal; say, $n_j = n$. Indeed, the sequence of numbers n_j is bounded in view of condition 4) of Theorem 2.8. Therefore the sequence (2) may be replaced by the subsequence involving those indices j for which corresponding numbers n_j are equal to n.

Depending on C we may choose α_0 sufficiently small such that, by condition 3) of Theorem 2.8, the inequality $n \geqslant 2$ holds.

We denote by L_j a hyperplane for which $L_j \cap Z_n$ is the kernel φ_j; by L_{m_j} we denote the hyperplane for which $(L_{m_j} \cap Z_n)\varphi_j = m$. Let $a_{k_{j-1}}$ be the largest of numbers $a_i \leqslant x_j/2$. Then we construct a cylinder N_j which extends one of the edges of the parallelepiped H_j (the method of choosing this edge will be described below).

Let F be the strip between the hyperplanes L_j and $L_{a_{k_j-1}j}$; furthermore let $H'_j = N_j \cap F$. Then the remark following Lemma 2.13 implies that we may choose the edge of the parallelepiped H_j in such a way that

$$V(H'_j) \leqslant nV(H_j).$$

For any positive constant c_1 we may assume that the set cP'_j with $P'_j = N_j \cap L_j$

does not contain any two points which are congruent mod Γ_j ($\Gamma_j = L_j \cap Z_n$). This follows from the fact that in the conclusion of the proof of Theorem 2.8 we may apply Lemma 2.27 instead of Lemma 2.26 in 2.8. In this case the constant c of condition 3) in Theorem 2.8 depends on c_1.

Applying Lemma 3.5 we find a basis $\bar{a}_{1j}, \ldots, \bar{a}_{n-1j}$ for the lattice Γ_j such that P_j' is contained in the parallelepiped with its center at the origin and with edges $c_2\bar{a}_{1j}, \ldots, c_2\bar{a}_{n-1j}$, where c_2 is a positive constant which may be chosen arbitrarily small.

A linear transformation $\varphi_j' : E_n \to E_n$ is defined in the following manner: Let $\bar{a}_{ij}\varphi_j' = \bar{e}_i$, $1 \leqslant i \leqslant n - 1$. A vector $\bar{a}_{nj} \in L_{1j}$ with $\bar{a}_{nj}\varphi_j' = \bar{e}_n$ is defined such that the parallelepiped $H_j'' = H_j\varphi_j'$ satisfies condition (2.7.2) of the definition of a canonical parallelepiped. Then H_j'' has height $h_{nj} = a_{k_j - 1}$ and a base P_j'' contained in the square P' with sides $c_2\bar{e}_1, \ldots, c_2\bar{e}_{n-1}$ and with its center at the origin. A sequence of edges of the parallelepiped H_j'' which do not lie in $L(\bar{e}_1, \ldots, \bar{e}_{n-1})$ may be assumed to converge in direction to some direction l which is not parallel to $L(\bar{e}_1, \ldots, \bar{e}_{n-1})$. Indeed, from the sequence of edges under consideration we may choose a subsequence which is convergent in direction, and then the sequence (2) may be taken to be this subsequence, changing indices appropriately. Condition (2.7.2) guarantees that the limiting direction l is not parallel to $L(\bar{e}_1, \ldots, \bar{e}_{n-1})$.

We consider the cylinder N' with base P' and l as generator. The sets $N' \cap Z_n$ and $(N' \cap Z_n)\varphi$ are isomorphic. The set N' contains at least one inverse image of each of the numbers a_i (i.e. $N' \cap a_i\varphi^{-1} \neq \varnothing$). Indeed, $N_j'' \cap a_i\varphi^{-1} \neq \varnothing$ for sufficiently large j, and the set N_j'' to be mapped converges to l in direction.

For the numbers a_i contained in N' we consider the projection of the inverse images $a_i\varphi^{-1}$ onto $L(\bar{e}_1, \ldots, \bar{e}_{n-1})$ parallel to l. Let P be the convex hull of this projection.

We show that $V(P) < c\alpha$, where c is a positive constant depending on C and C_1. For sufficiently large k, the smallest convex polyhedron P'' containing the projection of $a_0, \ldots, a_{k-1}$ has a volume which is arbitrarily close to $V(P)$. But for sufficiently large j we have $V(P) < (1 + \varepsilon)V(P_j'')$, where ε is a positive constant to be chosen arbitrarily small, and $P_j'' = N_j'' \cap L(\bar{e}_1, \ldots, \bar{e}_{n-1})$. Inasmuch as $V(H_j'') < c_3 k_j$ this implies $V(P_j'') < 3^k j / a_{k_j - 1} < c_4 \alpha$.

Finally, the condition $T(N^{(x)} \cap Z_n) < c\alpha x$ follows from the inequality $V(H_j'') < c_3 k_j$ and from the fact that $a_i\varphi^{-1} \cap N'$ coincides with $a_i\varphi^{-1} \cap N_j''$ for sufficiently large j in all cases where the latter intersection is nonempty. But this condition follows from (1) if we assume that c_2 is sufficiently small in relation to C_1.

COROLLARY 1. *Under the conditions of Theorem 3.6*

$$\varlimsup_{x \to \infty} \frac{A(x)}{x} < c\alpha.$$

COROLLARY 2. *Under the conditions of Theorem 3.6 the order r of the basis of A satisfies the inequality $r > c/\alpha^{1/n}$.*

3.7. *Analysis of the case $C < 2$.* The special case $C < 2$ of the problem considered in §3.6 was studied by M. Kneser [27]. In this subsection we show that this theorem follows from our general results, and in §3.8 we discuss the connection with Kneser's results.

THEOREM. *Let A be a sequence with $a_0 = 0$, $d(A) = 1$, $\alpha > 0$, $\gamma < C\alpha$ and $3/2 \leqslant C < 2$. Then there exists a positive number $\alpha_0(C)$ and positive integers g, h, l, $h_0 = h_0(C)$ and u_0 with $(g, l) = 1$ such that $h < h_0$ and for $\alpha < \alpha_0$ the sequence A is contained in the union of h residue classes modulo g congruent to the residues which are contained in the arithmetical progression*

$$a_0, a_0 + l, a_0 + 2l, \ldots, a_0 + (h - 1)l; \tag{3.7.1}$$

furthermore,

$$\sigma > \kappa h/g, \tag{3.7.2}$$

where $\kappa > (h - 1)/4h$, and, in particular,

$$\kappa > 1/2 \tag{3.7.3}$$

if $h = 2$.

PROOF. We consider the sequence of sets K_j obtained in the proof of Theorem 3.6. Instead of this theorem we have further to apply Theorem 2.8 if $C < 2$ which is, in view of (6.6), applicable to all K_j with sufficiently large j.

We obtain a sequence of polygons H_j, containing the origin, whose sides are parallel to the coordinate axes and with the property that the set of lengths of the edges parallel to the coordinate axes is uniformly bounded by a constant c depending only on C.

Let $(1, 0)\varphi_j = g_j$. The numbers g_j, which may be assumed to be positive, are uniformly bounded since the interval $[0, g_j]$ consists in a unique manner of the sets of integers contained in the intersection of H_j with any straight line parallel to the horizontal axis, and there are no more than c such lines whose intersection with H_j is nonempty. By choosing the corresponding subsequence we may therefore reduce our consideration to the case where $g_j = g$, and then to the case where all points from $(0, 1)\varphi_j$ belong to the same residue class modulo g. Finally, we may assume that all points $f_j = (0, 1)\varphi_j$ are identical and belong, for example, to the interval from 0 to g, i.e. $f_j = f$, $0 < f < g$. This identity may be attained by changing each φ_j by a shift parallel to the horizontal axis.

Thus all φ_j may be assumed to be identical.

As a result we obtain a strip F containing the lines $x_2 = b$, $-t_1 \leqslant b \leqslant t_2$, where t_1 and t_2 are given nonnegative numbers, and a homomorphism $\varphi: Z_2 \to Z_1$ such that $A \subset (F \cap Z_2)\varphi$ and $F \cap Z_2$ are isomorphic to $(F \cap Z_2)\varphi$.

In order to complete the proof of the theorem it remains to establish the inequality (2).

For this purpose we determine the precise structure of A. Let $\{b_i\}$ be the set of numbers b_i, ordered by increasing magnitude, for which

$$A \cap \{(x_2 = b_i) \cap Z_2\}\varphi \neq \varnothing .$$

Let $M^{(x)}$ consist of those points x_1 in the set M for which $x_1 < x$. Let $D_i = A\varphi^{-1} \cap F^{(x)} \cap \{x_2 = b_i\}$, $T(D_i) = \delta_i$. We assume that for all sufficiently large x the inequality

$$T\left(\left(2\bigcup_{i>i_1}^{i_2} D_i\right)^{(x)}\right) \geqslant 4(\delta_{i_1} + \ldots + \delta_{i_2}) - c \tag{3.7.4}$$

is satisfied, where c as well as $c_1, \ldots, c_5$ below are sufficiently large positive constants. Then it is immediately evident that

$$T\left(\left(2\bigcup_{i>i_1}^{i_2+1} D_i\right)^{(x)}\right) \geqslant 4(\delta_{i_1} + \ldots + \delta_{i_2} + \delta_{i_2+1}) - c_1 .$$

Indeed,

$$T\left(\left(2\bigcup_{i\geqslant i_1}^{i_2+1} D_i\right)^{(x)}\right) \geqslant T((2D_{i_2+1})^{(x)}) + T((D_{i_2} + D_{i_2+1})^{(x)})$$

$$+ T\left(\left(4\bigcup_{i\geqslant i_1}^{i_2} D_i\right)^{(x)}\right)$$

$$\geqslant 2\delta_{i_2+1} - c_3 + 2(\delta i_1 + \ldots + \delta_{i_2}) - c.$$

Here we use the fact that if $\bar{z}_1$ and $\bar{z}_2$ are points with minimal abscissas belonging to the sets D_{i_2+1} and D_{i_2}, respectively, then

$$\bar{z}_1 + D_{i_2+1} \subset 2D_{i_2+1} \quad \text{and} \quad \bar{z}_2 + D_{i_2+1} \subset D_{i_2} + D_{i_2+1} .$$

Thus the inequality (4) contradicts the assumption $C < 2$.

We may assume that the numbers b_i are relatively prime. This may be accomplished by a contraction in the vertical direction. We show that all b with $-t_1 \leqslant b \leqslant t_2$ belong to $\{b_i\}$. We may assume, of course, that $-t_1, t_2 \in \{b_i\}$, since otherwise a certain number of straight lines $x_2 = b$, $b \notin \{b_i\}$, may be thrown out, thus narrowing the strip F.

If we assume that there exists $b \notin \{b_i\}$, $-t_1 < b < t_2$, then there exists an index i_1 such that $b_{i_1} + b_{i_1+2} \neq 2b_{i_1+1}$. The sets

$$D_j + D_t, \tag{3.7.5}$$

do not overlap with each other if j and t assume the following pairs of values: $j = t = i$; $j = t = i_1 + 1$; $j = t - i_1 + 2$; $j = i_1$, $t = i_1 + 1$; $j = i_1$, $t = i_1 + 2$; $j = i_1 + 1$, $t = i_1 + 2$. Thus the inequality (4) holds for the set $\bigcup_{i=i_1}^{i_1+2} D_i$.

We denote by q_i the greatest common divisor of the mutually different numbers of the sequence $A_i = A \cap (\{x_2 = b_i\} \cap Z_2)\varphi$. The quantities g and l of the theorem are defined by the equations $(1,0)\varphi = g$ and $(0,1)\varphi = l$. It is evident that $(l, g) = 1$ since $d(A) = 1$ and $h < h_0$ if α_0 is sufficiently small.

We show that

$$q = (q_1, q_2, \dots) = g. \tag{3.7.6}$$

Indeed, let $q > g$. If we assume that i_1 is such that $a_{i_1} + a_{i_1+2} \not\equiv 2a_{i_1+1} \pmod{q}$, where $a_j \in A_j, j = i_1, i_1 + 1, i_1 + 2$, then we can show as in the case of (5) that the six sets $D_j + D_t$ are mutually disjoint and that inequality (4) holds for the set $\bigcup_{i=i_1}^{i_1+2} D_i$.

It can be shown that $(0,1)\varphi$ is congruent modulo q with the numbers $A \cap (\{x_2 = 1\} \cap Z_2)\varphi$, and that the number $(b, 1)$ is congruent with the numbers $A \cap (\{x_2 = b\} \cap Z_2)\varphi$ modulo q. Clearly equation (6) may be satisfied by contracting the axis of abscissas in the ratio q/g.

Suppose K is an ordered set in a plane whose points have ordinates between $-t_2$ and $-t_2 + h - 1$; furthermore, let $(0,j) \in K$ for $-t_2 \leqslant j \leqslant -t_2 + h - 1$ and assume the abscissas of the points in K to be nonnegative, the greatest of them being equal to p. It may be assumed that $(p, 0) \in K$. We project the set K onto the straight line $x_2 = 0$ parallel to the vector $(-p, 1)$. The points on the axis of ordinates are projected onto the points pr with $-t_2 \leqslant r \leqslant t_2 + h - 1$, and the segment to which the projection belongs must contain at least $(h - 1)p + 1$ points.

If $T < 4k - c_4$, then we show as in §1.17 (see Exercise 5 there) that

$$(h - 1)p < T - 2k + c_5. \tag{3.7.7}$$

We denote by the largest of the numbers a_i not exceeding $y_j/2$. If the inequality $p < y_j/4g$ were to hold, then we would have

$$A_2(y_j) \geqslant T(2\{a_i \leqslant y_j/2\}) + A(y_j) - A(y_j/2)$$
$$\geqslant 3A(y_j/2) + A(y_j) - A(y_j/2)$$
$$\geqslant (\alpha - \varepsilon)y_j + 2(\alpha - \varepsilon)y_j/2 = 2(\alpha - \varepsilon)y_j.$$

Thus $p \geqslant y_j/4g$ and, in view of (7),

$$(h - 1)y_j/4g < (\gamma + \varepsilon)y_j - 2(\alpha - \varepsilon)y_j/2 < \alpha y_j,$$

which implies the assertion (2).

As an example showing the possibility of sharper results we take up the case $h = 2$.

First let us assume that $\alpha < 1/2g$. Let z_1 and z_2 be the largest numbers from A in two arithmetic progressions with difference g containing A. Thus $z_1, z_2 \leqslant y_j/2$. Then we have (see again §1.17, Exercise 5)

$$(z_1 + z_2)/g < (\gamma - \alpha + \varepsilon)y_j < \alpha y_j \leqslant y_j/2g.$$

If $z_1 \leqslant z_2$, then $2z_1 \leqslant y_j/2$,

$$2\left\{a_i < \frac{y_j}{2}\right\} \cap \left\{\frac{y_j}{2} < a_i < y_j\right\} = \varnothing$$

and

$$(\{a_i \in A, |a_i < y_j/2\} + \{a_i \in A_2 | a_i < y_j/2\})$$
$$\cap \{a_i + a_j | a_i \in A_2, a_j \in A_1, y_j/2 < a_i < y_j\} \neq \varnothing .$$

Therefore

$$A_2(y_j) \geqslant 3A(y_j/2) + A(y_j) - A(y_j/2) \geqslant 2(\alpha - \varepsilon)y_j \qquad (3.7.8)$$

and thus $\alpha > 1/2g$.

Let z_3 be the smallest of the numbers a_i for which $a_i \geqslant y_j/2$. For example, let $z_3 \in A_1$. Let $K = \{a_i \leqslant z_3\}$ and $k = T(K)$. Then

$$A(y_j/2) = A(z_3) - 1 > (\alpha - \varepsilon)z_3 .$$

We assume that $T(2K) \geqslant 4k - 6$. Thus

$$A_2(y_j) \geqslant 4k - \frac{2(z_3 - y_j/2)}{g} - 8,$$

and hence

$$(\gamma + \varepsilon)y_j \geqslant 4(\alpha - \varepsilon)z_3 - (2z_3 - y_j)/g.$$

This implies

$$(\gamma + \varepsilon - 2\alpha + 2\varepsilon)y_j \geqslant (2z_3 - y_j)(2\alpha - 2\varepsilon - 1/g),$$

which is impossible in view of (8).

Let $T < 4k - 6$. Then

$$(z_2 + z_3)/g < (\gamma + \varepsilon)y_j - 2(\alpha - \varepsilon)z_3.$$

If $z_2 + z_3 \geqslant y_j$, this implies $\alpha > 1/g$.

Now let $z_2 + z_3 < y_j$. We consider the set $K = \{a_i < y_j - z_2\}$. First let $T \geqslant 4k - 6$. Then

$$A_2(y_j) \geqslant 4(\alpha - \varepsilon)(y_j - z_2) - \frac{2(y_j/2 - z_2)}{g}(\gamma + 3\varepsilon - 2\alpha)y_j$$

$$\geqslant (y_j - 2z_2)\left(2\alpha - 2\varepsilon - \frac{1}{g}\right).$$

If $T < 4k - 6$, then

$$(\gamma + \varepsilon)y_j - 2(\alpha - \varepsilon)(y_j - z_2) \geqslant (y_j/2 + z_2)/g,$$

and thus

$$(\gamma + 2\varepsilon - \alpha - 1/g)y_j \geqslant (y_j/2 - z_2)(2\alpha - 2\varepsilon - 1/g).$$

Hence the inequality $\alpha > 1/g$ follows by means of (8).

3.8. *Connection with Kneser's theorem.* The same structure of a set A is described by Theorem 3.7 and by Theorem 3.2. It is shown in either case that A is contained in the union of a bounded number of residue classes. In Theorem 3.7 the mutual position of these residue classes is described in greater detail. This is not done in Theorem 3.2 (in this direction a more precise result was obtained by Kemperman [28]). However, the lower bound for α in Theorem 3.7 is considerably weaker than that which was proved in (2.2).

The difficulties in obtaining the best possible lower bound for α are caused by the fact, which we have already observed, that the method of trigonometric sums may be successfully applied only for investigating the structure of sufficiently "thin" sets. Again we have to underline that the chief value of the methods developed in this book lies in the possibility of studying the structure of sets with arbitrary values of C, whereas the earlier methods were not suitable for the case $C \geqslant 2$.

For the case $h = 2$ we have obtained the estimate $\alpha > 1/g$. It is very likely that the inequality

$$\alpha > (h - 1)/g \tag{3.8.1}$$

is true in Theorem 3.7. This inequality would imply (7.3) as a special case ($h = 2$).

For this purpose it is necessary to consider the case $T < Ck$, $C < 4$, as was done for $T < 10k/3 - 5$ in §1.17.

We wish to make still another observation concerning the relation between

Theorems 3.2 and 3.7, assuming inequality (1) in the latter theorem to be proved completely. It follows from (1) that for any residue class modulo g whose elements are congruent to one of the numbers

$$2u_0 + l, 2u_0 + 2l, \ldots, 2u_0 + (2h - 3)l,$$

all sufficiently large elements belong to $2A$.

If the residue classes have the structure described in §3.7, then inequality (1) is best possible. Indeed, let us consider as an example the set containing all positive numbers which are congruent modulo g with any of the first $h - 1$ numbers in (7.1) and congruent modulo tg with the number $a_0 + (h - 1)l$.

In Theorem 3.2 no stipulation is made concerning the existence of a structure of the residue classes defined by the numbers (7.1). In the example mentioned we may therefore consider all residue classes modulo tg which contain numbers of the set. There are $t(h - 1) + 1$ such residue classes, whereas the set $2A$ contains $2t(h - 1) + 1$ residue classes and the inequality (2.2) is satisfied with respect to these classes.

A more precise description of the structure of A by means of the inequality (2.2.1) obtained above would be related to an investigation of the structure of A for large values of α. The results developed in this book do not imply such an assertion.

§3. Sums of sets with asymptotic density zero

3.9. *An inverse additive theorem for sets of asymptotic density zero.*

THEOREM. *Suppose there exists a constant $C > 0$ such that*

$$\delta = \varlimsup_{x \to \infty} \frac{A_2(x)}{A(x)}; \quad \alpha = \lim_{x \to \infty} \frac{A(x)}{x} = 0.$$

Then there exist infinitely many $a_i \in A$ for which $2a_i < a_{i+1}$.

PROOF. If there existed only finitely many k such that $2A(a_{k-1}) = 2k \geqslant A(2a_{k-1})$, then there would exist a sufficiently large u for which $A(2^s a_{u-1}) \geqslant 2^s u$ for any positive integer s, and this would imply the inequality $\alpha > 0$. But since $\alpha = 0$ by assumption, there exist infinitely many numbers k satisfying $2A(a_{k-1}) \geqslant A(2a_{k-1})$. For sufficiently large ones among these k we have

$$T(2K) \leqslant A_2(2a_{k-1})$$

$$= (A_2(2a_{k-1})/A(2a_{k-1}))A(2a_{k-1}) \leqslant (C + \varepsilon)2k = 2(C + \varepsilon)k.$$

To each of these K we may apply Theorem 2.8.

Now we choose numbers k and k_0 such that $\varepsilon = k_0/a_{k_0-1}$ is sufficiently small; furthermore we let $\varepsilon_2 = V/M$, where $V = V(H)$ is the volume of the parallelepiped H defined by applying Theorem 2.8 to the set K, and $M = a_{k-1}$, with ε_2 sufficiently small in relation to ε_1 (the method of choosing these constants will be made clear in the sequel).

By Theorem 2.8 a homomorphic mapping $\varphi: Z_n \to Z_1$ is defined. Let $\bar{e}_i \varphi = g_i$, $1 \leqslant i \leqslant n$. We may assume that $g_i > 0$. Then $\bar{f}_i \varphi = u$, where $\bar{f}_i = u\bar{e}_i/g_i$ and $u = [g_1, \ldots, g_n]$.

For any real number μ we denote the hyperplane $\nu_1 \bar{f}_1 + \nu_2 \bar{f}_2 + \ldots + \nu_n \bar{f}_n$, $\sum_{i=1}^n \nu_i = \mu$, by D_μ. The set $D_0 \cap Z_n$ is the kernel of the homomorphism φ.

Let the cylinder N_i be defined as the sets of points

$$\bar{u} + \sigma_1 \bar{x}_1 + \sigma_2 \bar{x}_2 + \ldots + \sigma_n \bar{x}_n, \qquad -\infty < \sigma_i < \infty \qquad (3.9.1)$$

$$0 \leqslant \gamma_j < 1, 1 \leqslant j \leqslant n, j \neq i,$$

where $\bar{u}$ is the height of the parallelepiped H.

By F_μ we denote the closed strip between the hyperplanes D_0 and D_μ. Let $D_{\bar\mu}$ be that supporting hyperplane of H parallel to D_0 which lies on the same side of D_0 for which the images of the integral points are positive. Furthermore let $W_\mu = N_i \cap F_\mu$.

It follows from the remark to Lemma 2.13 that a number i may be defined in such a way that the volume of $W_{\bar\mu}$ does not exceed nV. Therefore the segment $\lambda_{\bar\mu} \bar{x}_i$ on the line $\lambda \bar{x}_i$, $-\infty < \lambda < +\infty$, between the hyperplanes D_0 and $D_{\bar\mu}$ satisfies the condition $\lambda_{\bar\mu} \leqslant n$. We show that for sufficiently small ε_1 and ε_2 there exist numbers $\lambda_{\mu'}$ and $\lambda_{\mu''}$ for which $\lambda_{\mu'}/\lambda_{\mu_0} \geqslant 1$, $\lambda_{\mu_0} = a_{k_0-1}/u$ and $\lambda_{\mu''}/\lambda_{\mu'} > 2$, such that $Z_n \cap (W_{\mu''} \setminus W_{\mu'}) = \varnothing$. Since this shows that $A \cap [u\lambda_{\mu'}/\lambda_u, u\lambda_{\mu''}/\lambda_u] = \varnothing$, the assertion of the theorem follows immediately.

The dimension of the set $Z_n \cap W_\mu$ will be denoted by p_μ. A method of constructing the numbers

$$\mu_1, \bar\mu_1, \mu_2, \bar\mu_2, \ldots, \mu_{i_0}, \bar\mu_{i_0} \qquad (3.9.2)$$

will now be defined inductively. We let $\mu_1 = \mu_0$. Suppose the numbers (μ_i) have already been constructed up to a given i. We consider the set of real numbers μ for which $\lambda_\mu/\lambda_{\bar\mu} > 0$, $p_\mu = p_{\mu_i}$ and $T(W_\mu \cap D_\mu) > 0$. By $\bar\mu_i$ we denote the supremum or infimum of the set of numbers μ if they are positive or negative, respectively. Now we consider the set of real numbers μ for which $\lambda_\mu/\lambda_{\bar\mu} > 0$ and $W_\mu \setminus (W_{\bar\mu_j} \cup D_\mu) = \varnothing$. Let μ_{i+1} be the supremum or infimum of the numbers μ if they are positive or negative, respectively. We continue this construction up to an index i_0 for which $p_{\mu_{i_0}} = n$. We define $\bar\mu_{i_0}$ by the relation $\lambda_{\bar\mu_{i_0}} = \lambda_{\bar\mu}$. Then we select within W_{μ_i} a set of dimension p_{μ_i} consisting of $p_{\mu_i} + 1$ points. Let H_i be the parallelepiped with these points as vertices. By U_1 we denote the cylinder defined in (1), with $-2n \leqslant \sigma_j < 2n + 1$, $j \neq i$, and $W_{1\mu} = U_1 \cap F_\mu$.

The projection of the points of the parallelepiped H_i onto $\lambda \bar{x}_i$ parallel to D_0

defines a segment whose length does not exceed $n|\lambda_{\mu_i}|\,\|\bar{x}_i|$. In any segment of length $n|\lambda_{\mu_i}|\,\|\bar{x}_i|$ contained in the segment $\lambda_{\bar{\mu}_i}$ there exist the projections (parallel to D_0) of integral vectors contained in W_{μ_i}. Indeed, we observe first that all integral points contained in W_{μ_i} (including those belonging to $D_{\bar{\mu}}$) are contained in the linear space generated by the vertices of H_i. Secondly, any parallelepiped which is obtained within this linear subspace by a parallel shift of H_i will always contain a lattice point. Thus, in the notation

$$k_i = T(W_{\mu_i} \cap Z_n), \quad \bar{k}_i = T(W_{1\bar{\mu}_i} \cap Z_n), \quad i = 1, 2, \ldots, i_0,$$

the inequality

$$\bar{k}_i \geqslant (\lambda_{\bar{\mu}_i}/4n\lambda_{\mu_i})k_i, \quad i = 1, 2, \ldots, i_0, \tag{3.9.3}$$

holds.

We assume that

$$\lambda_{\mu_{i+1}}/\lambda_{\bar{\mu}_i} \leqslant 2, \quad i = 1, 2, \ldots, i_0 - 1. \tag{3.9.4}$$

Then there exists a constant $c = c(n) > 0$ such that

$$\bar{k}_i < ck_{i+1}, \quad i = 1, 2, \ldots, i_0 - 1. \tag{3.9.5}$$

From (3), (4) and (5) we obtain the inequality

$$\bar{k}_{i_0} \geqslant \frac{\lambda_{\bar{\mu}_{i_0}}}{\lambda_{\mu_1}} \frac{1}{(4n)^{i_0}(2c)^{i_0-1}} k_1.$$

But $k_1 = k_0$; furthermore,

$$\bar{k}_{i_0} < c_1(n)V,$$

and

$$\lambda_{\bar{\mu}_{i_0}}/\lambda_{\mu_1} \geqslant M/a_{k_0-1}.$$

Therefore

$$\varepsilon_2 \geqslant C_2\varepsilon_1, \tag{3.9.6}$$

where

$$C_2 = 1/(4n)^{i_0-1}(2c)^{i_0-1}c_1(n).$$

For any sufficiently small $\varepsilon_1 > 0$ there exists an ε_2 such that the inequality (6) is violated. Hence the assumption (4) is not satisfied, and there exists an index i such that $\lambda_{\mu_{i+1}}/\lambda_{\bar{\mu}_i} > 2$. This proves the theroem.

COROLLARY. *If $\alpha = 0$ and if the inequality $a_{i+1} > 2a_i$ is satisfied for a finite number of indices only, then*

$$\overline{\lim_{x \to \infty}} \frac{A_2(x)}{A(x)} = +\infty.$$

3.10. *Proof of a conjecture of P. Erdös.* In [10] (Problem 15) P. Erdös stated the conjecture that

$$\delta = \overline{\lim_{x \to \infty}} \frac{A_2(x)}{A(x)} \geqslant 3 \qquad \text{if } \lim_{x \to \infty} \frac{A(x)}{x} = 0.$$

The following somewhat sharper result is true:

THEOREM. *If $\alpha = 0$ and $\overline{\lim}_{x \to \infty} A(x)/x < 1/2d(A)$, then $\delta \geqslant 3$.*

PROOF. Suppose $\delta < 3$. We consider the sequence of those k_i, $i = 1, 2, \ldots$, for which $a_{k_i} > 2a_{k_i-1}$. If the relation $T(2K_i) \geqslant 3k_i - 3$ were true for infinitely many k_i, then we would have $\delta \geqslant 3$ since

$$A_2(2a_{k_i-1})/A(2a_{k_i-1}) \geqslant (3k_i - 3)/k_i.$$

Neglecting a finite number of terms in the sequence k_i if necessary, we may thus always assume $T(2K_i) < 3k_i - 3$. For sufficiently large k_i it follows from Theorem 1.9 that $a_{k_i} \leqslant (2k_i - 4)d(A)$ and hence

$$\overline{\lim_{x \to \infty}} \frac{A(x)}{x} \geqslant \frac{1}{2d(A)},$$

which proves the theorem.

3.11. *The structure of A for $\delta < 10/3$.* The result of the preceding subsection may be sharpened as follows:

THEOREM. *If $\delta < 10/3$, $d(A) = 1$ and $\lim_{x \to \infty} A(x)/x = 0$, then there exists a sequence $i_1 < i_2 < \ldots$ of positive integers such that*

$$\overline{\lim_{s \to \infty}} \frac{a_{i_s} + a_{i_{s+1}} - a_{i_s+1}}{A(a_{i_{s+1}})} \leqslant \delta - 2.$$

PROOF. We consider the sequence of all indices i_s, $s = 1, 2, \ldots$, for which $a_{i_s+1} > 2a_{i_s}$, with i_s greater than a positive constant i_0 to be chosen sufficiently large. If there existed infinitely many s with

$$T(2K_s) \geqslant \frac{10}{3}(i_s + 1) - 5,$$

where $K_s = \{a_0, a_1, \ldots, a_{i_s}\}$, then it would follow that $\delta \geqslant 10/3$. Therefore we may always assume

$$T(2K_s) < (10/3)(i_s + 1) - 5.$$

Without loss of generality we let $d(K_s) = 1$. By Theorem 1.17 the sets K_s are contained in two arithmetic progressions with identical differences d_s whose length does not exceed $i_s + 1 + b_s$, where b_s is defined by the equation $T(2K_s) = 3i_s + b_s$. Consequently the segment $[0, d_s - 1]$ does not contain more than two points from A, and hence d_s cannot assume an unbounded range of values. Suppose there exists an infinite set of numbers $d_s > 1$; then we consider a subsequence $\{K_{s_t}\}$ for which $d_{s_t} = d > 1$ $(T = 1, 2, \ldots)$. Inasmuch as $d(A) = 1$, there exist numbers of either of the two progressions with distance d within a bounded distance from the origin. Hence $T(K_{s_t}) > ca_{i_{s_t}}/d$, which contradicts the condition $\lim_{x \to \infty} A(x)/x = 0$. Thus we may assume $d_s = 1$ for all s. This proves the theorem.

§4. Sums of sets of residues modulo a prime

3.12. *The structure of sets of residues with small double set.* Some results in this direction have been obtained in §1 of Chapter II. For any positive constant C it is possible to investigate the case $T < Ck$ by means of Theorem 3.6.

THEOREM. *Let K be a set consisting of k residues modulo the prime number p. If $T < Ck$, $C \geqslant 2$, then there exist positive constants c and c_1, depending only on C, a homomorphism $\varphi\colon Z_n \to S_p$, where S_p is the additive group of residues modulo the prime number p, and a canonical parallelepiped $H \subset E_n$ such that the following assertions are true for $k < cp$:*
1) $K \subset (H \cap Z_n)\varphi$,
2) $H \cap Z_n$ and $(H \cap Z_n)\varphi$ are isomorphic,
3) $T(H \cap Z_n) < c_1 k$,
4) $n \leqslant [C - 1]$.

PROOF. We consider the set consisting of all nonnegative numbers in those residue classes modulo p which contain residues from the set K. It is evident that this set satisfies $\alpha = k/p$ and $\gamma = T/k < Ck/p$. Thus Theorem 3.6 is applicable. It is obvious that the generating line l of the cylinder N in Theorem 3.6 runs through a set of lattice points which is mapped onto a class of residues modulo p. We consider a basis $\bar{a}_1, \ldots, \bar{a}_n$ of the lattice Z_n such that the vector $\bar{a}_n$ is parallel to l and satisfies $\bar{a}_n \varphi = p$.

The set $P_1 = N \cap L(\bar{a}_1, \bar{a}_2, \ldots, \bar{a}_{n-1})$ is convex. It may be embedded in the canonical parallelepiped H' relative to the basis $\bar{a}_1, \ldots, \bar{a}_{n-1}$. The homomorphism induced by φ on $L(\bar{a}_1, \ldots, \bar{a}_{n-1})$ and the parallelepiped H' have the desired properties.

COROLLARY. *If $k < cp$, where $c > 0$ is a sufficiently small positive absolute constant, and if $T < 3k - 3$, then K is contained in a progression modulo p of length $k + b$, where $b = T - 2k + 1$.*

3.13. *Further considerations on sums of sets of residues modulo a prime.* The main result of Theorem 3.12 may be reformulated as follows: If $T < Ck$ and $k < cp$, $c = c(C) > 0$, then there exists an isomorphic mapping of the set K into a set of integers.

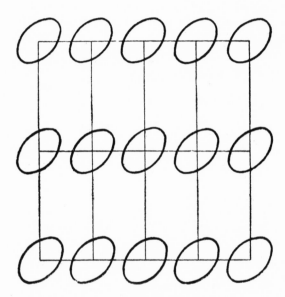

Figure 6

It is known that sets of integers and sets of residues are different if considered from an additive viewpoint: addition of sets of integers takes place on "the line" whereas addition of sets of residues is performed "on the circle". However in the case of noniterated addition of sets of residues which are not too large, this circle is "broken up". This phenomenon expresses itself by the fact that such a set of residues is isomorphic to the corresponding set of integers.

If $k > p/2 + 1$, then $T < 2k - 1$, and there is no explicit isomorphism.

If $k < p/2$, then Vosper's theorem implies that for $T = 2k - 1$ the set of residues is isomorphic to an arithmetic progression consisting of k integers.

We may assume that for a pair k and T satisfying

$$W(1, T, k) < p/2,$$

where $W(1, T, k)$ is defined by (1.24.1), the set of residues under consideration may be mapped isomorphically onto some set of integers. Thus the problem of adding sets of residues modulo a prime can be reduced to the problem of adding sets of integers.

§5. Sums of point sets

3.14. *Sums of sets of positive measure.*

THEOREM. *Let* $\mu^*(2D) < C\mu^*(D)$, $C \geqslant 2^m$, *and let* D *be a set in* E_m *with positive outer Lesbesgue measure* $\mu^*(D)$. *Then there exist a homomorphism* $\varphi\colon Z_n \to E_m$, *a rectangular canonical parallelepiped* $H \subset E_n$ *and a convex set* $D_1 \subset E_n$ *such that the following assertions are true:*
1) $D \subset (H \cap Z_n)\varphi + D_1$,
2) $(H \cap Z_n) \times D_1 \subset E_n \times E_m$ *and* $(H \cap Z_n)\varphi + D_1$ *are isomorphic,*
3) $T(H \cap Z_n)\mu(D_1) < c\mu^*(D)$, $c = c(C, m)$,
4) $n \leqslant [C - 2^m]$.

PROOF. We subdivide E_m into m-dimensional cubes of rank p.[3] In order to accomplish this we consider m systems of hyperplanes

$$x_j = 0, \pm 1, \pm 2, \pm 3, \ldots, 1 \leqslant j \leqslant m.$$

These hyperplanes subdivide the space E_m into a denumerable family of cubes, each of them considered with its boundary, with edges of length one. The interiors of these cubes are mutually disjoint. We call them cubes of rank one.

Furthermore we introduce hyperplanes of the form

$$x_j = 0, \pm\tfrac{1}{2}, \pm 1, \pm\tfrac{3}{2}, \ldots, 1 \leqslant j \leqslant m.$$

These hyperplanes subdivide the space E_m into cubes which we call cubes of rank two. It is evident that each cube of rank one consists of 2^m cubes of rank two.

As the next step, we consider systems of hyperplanes of the form

$$x_j = s_j/4, s_j = 0, \pm 1, \pm 2, \ldots, 1 \leqslant j \leqslant m,$$

$$x_j = s_j/8, s_j = 0, \pm 1, \pm 2, \ldots, 1 \leqslant j \leqslant m.$$

---

[3] See [34], Chapter XI, §2.

Continuing this process indefinitely, we obtain cubes of ranks 3, 4, etc.

All these cubes are closed, and their faces are parallel to the coordinate hyperplanes. Two cubes of the same rank do not have any interior points in common; each cube of rank p consists of 2^m cubes of rank $p + 1$, and the edges of a cube of rank p have length 2^{1-p}. Finally, the set of all these cubes is denumerable.

Let M be a set in E_m. We consider the set of all entire cubes of rank p which intersect with M. The union of all these cubes will be denoted by $U_p(M)$. By $W_p(M)$ we denote the set of vertices of the cubes of rank p which are contained in a given set M. We consider the set of all those cubes which are obtainable from cubes of rank p by parallel shifts and which possess at least one vertex within a given set M. The union of all these cubes is denoted by $Q_p(M)$.

For a given set $M \subset E_m$ we determine an open set $G \supset M$ such that $\mu(G) < \mu^*(M) + \varepsilon$, where ε is a given positive number. For any $\varepsilon_1 > 0$ there exists a sufficiently large number p satisfying

$$\mu(U_p(G)) < \mu(G) + \varepsilon_1. \tag{3.14.1}$$

Let R be a cube of rank p. We show that for any $\varepsilon_2 > 0$ there exists a $p_1 > p$ such that

$$\mu(Q_{p_1}(W_{p_1}(R))) < \mu(R)(1 + \varepsilon_2). \tag{3.14.2}$$

Indeed, there are $(2p_1 - 2p + 1)^m$ vertices of squares of rank $p_1 > p$ contained in R. Therefore

$$\mu(Q_{p_1}(W_{p_1}(R))) = (2p_1 - 2p + 2)^m 2^{m(1-p_1)}.$$

Since

$$\mu(R) = (2p_1 - 2p)^m 2^{m(1-p_1)},$$

the assertion (3.14.2) is satisfied for sufficiently large values of $p_1 - p$.

In view of (1) and (2), the inequality

$$\mu(Q_{p_1}(W_{p_1}(U_p(G)))) < \mu(G)(1 + \varepsilon_3)$$

is valid for any $\varepsilon_3 > 0$. Hence we have

$$\mu(Q_{p_1}(W_{p_1}(U_{p_1}(G)))) < \mu(G)(1 + \varepsilon_3). \tag{3.14.3}$$

We consider an open set $G_1 = 2D$ such that

$$\mu(G_1) < \mu^*(2D) + \varepsilon_4.$$

Furthermore, let p be chosen in such a way that

$$\mu(U_p(G_1)) < \mu(G_1) + \varepsilon_5.$$

By (3) the set $Q_p(W_p(U_p(G_1)))$ satisfies the inequality

$$\mu(Q_p(W_p(U_p(G_1)))) < \mu^*(2D)(1 + \varepsilon_6).$$

It is evident that $2U_p(D) \subset Q_p(W_p(U_p(G_1)))$, inasmuch as any cube with edges of length $2 \cdot 2^{1-p}$ containing points of $2D$ must belong to Q_p, and the sum-set of any two cubes from U_p which both contain points from D will itself be a cube containing points from $2D$. Analagously,

$$2Q_p(W_p(U_p(D))) \subset Q_{p-1}(W_{p-1}(U_{p-1}(2D))).$$

We use the notation $K = W_p(U_p(D))$.
It is evident that

$$T \leqslant (1/2^{1-p})\mu(Q_{p-1}(W_{p-1}(U_{p-1}(2D)))).$$

Since $T < (C + \varepsilon)k$, Theorem 2.8 is applicable to the set K. We obtain a mapping $\varphi_0 : Z_{n_1} \to 2^{1-p}Z_m$ and a canonical parallelepiped H_1 for which the conditions of the theorem are satisfied.

We may assume that $0 \in D$. Hence there exists a p_0 such that the set G_1 contains a square of rank p_0 such that the coordinates of its vertices are nonnegative numbers.

We determine vectors $\bar{a}_1, \ldots, \bar{a}_m \in E_{n_1}$ such that $\bar{a}_i\varphi_0 = 2^{1-p}\bar{e}_i$. We include additional vectors $\bar{a}_{m+1}, \ldots, \bar{a}_{n_1}$ in order to obtain a basis Z_{n_1}. We consider a linear transformation $\varphi_1 : Z_{n_1} \to Z_{n_1}$, defined by the relations $(2^{1-p}\bar{e}_i)\varphi_1 = \bar{a}_i$, $1 \leqslant i \leqslant m$, and $\bar{e}_i\varphi_1 = \bar{a}_i$, $i > m$. The transformation $\varphi_1\varphi_0$ defines a homomorphic mapping of the lattice $\Gamma_1 \subset E_{n_1}$ with basis $2^{1-p}\bar{e}_1, \ldots, 2^{1-p}\bar{e}_m, \bar{e}_{m+1}, \ldots, \bar{e}_{n_1}$ into the lattice $2^{1-p}Z_m$, where a point contained in $L(\bar{e}_1, \ldots, \bar{e}_m)$ is mapped onto the point $z_m \in Z_m$ with the same first m coordinates. Each of the mappings of the lattices induces a linear transformation of the corresponding euclidean spaces. The inverse image $H_1\varphi_1^{-1}$ is a certain convex set D'. Applying Lemma 2.13 we may embed D' in a parallelepiped H_2 satisfying condition (2.7.2). Since $h_j < c$, where c is a sufficiently large positive constant, the set H_2 may be embedded in a set H_3 for which the vector $\bar{x}_j$ is parallel to $\bar{e}_j$ if $m + 1 \leqslant j \leqslant n_1$. Of course, the volume is enlarged a bounded number of times by all these embeddings. The mapping $\varphi_1\varphi_0$ induces a mapping $\varphi_2 : (L(\bar{e}_{m+1}, \ldots, \bar{e}_{n_1}) \cap Z_{n_1}) \to Z_m$. We denote by H_4 the parallelepiped with edges $\bar{x}_{m+1}, \ldots, \bar{x}_{n_1}$ and with vertices in common with H_1. Then the sets

$$Z_n = L(\bar{e}_{m+1}, \ldots, \bar{e}_{n_1}) \cap Z_{n_1} \quad \text{and} \quad H = H_4$$

have the desired properties.

In order to show that $n \leqslant [C - 2^m]$ we consider the set of those points from the lattice Γ_3 which are contained in H_3. As in §1.17 we project this set onto $L(\bar{e}_1, \ldots, \bar{e}_{n_1-1})$, then onto $L(\bar{e}_1, \ldots, \bar{e}_{n_1-2}, \bar{e}_{n_1})$, up to $L(\bar{e}_1, \ldots, \bar{e}_m, \bar{e}_{m+2}, \ldots, \bar{e}_{n_1})$, each time choosing a vector in the direction of the projection such that within $L(\bar{e}_1, \ldots, \bar{e}_{n_1-1})$ there are but a finite number of points (no more than $\bar{h}_{n_1}$); hence the same is true for $L(\bar{e}_1, \ldots, \bar{e}_{n_1-2}, \bar{e}_{n_1})$, etc. Thus there are more than $k - c$ points contained in $L(\bar{e}_1, \ldots, \bar{e}_m)$, where c is a sufficiently large positive constant. Now it is evident that

$$T \geqslant (n_1 - m)k + 2^m k + o(k),$$

and therefore

$$n = n_1 = m \leqslant [C - 2^m].$$

It is evident that canonical parallelepipeds with equal volume have lattice points in common if their height is sufficiently large and if corresponding vertices are sufficiently close to each other. From this fact it may be easily derived that if $\bar{z}_1$ and $\bar{z}_2 \in H$, then the sets $\bar{z}_1 \varphi_2 + H_3 \varphi_1 \varphi_0$ and $\bar{z}_2 \varphi_2 + H_3 \varphi_1 \varphi_0$ are disjoint and their union contains D.

SUMMARY
ON GENERAL INVESTIGATIONS IN ADDITIVE NUMBER THEORY
Report delivered at the Number Theory Summer School in Palanga on September 5, 1965.

The title of this report has been chosen as an allusion to Ostmann's survey of additive number theory ([35], [36]), which was published in two volumes in 1955/56. The first volume is devoted to general investigations, i.e. to questions of sums of sets of positive density. The second volume contains a review of special problems such as Goldbach's problem, Waring's problem, the theory of partitions, etc.

1. Mann's theorem (1942) was the main result in the theory of density.

The Šnirel' man density of an increasing sequence (or set) $A = \{a_0, a_1, \ldots\}$ of nonnegative integers is defined as

$$\alpha'(A) = \inf_{x \in N} \frac{A(x)}{x},$$

where $A(x)$ is the number of positive elements in the set A which do not exceed x, and N is the set of natural numbers.

MANN'S THEOREM. *Let B and C be two sets of nonnegative integers with* $0 \in B$ *and* $0 \in C$. *Then*

$$\alpha'(B + C) \geqslant \min(1, \alpha'(B) + \alpha'(C)).$$

In recent years the evolution of the theory of density has been slow, due to the fact that both in methods and in results it essentially amounted to developments of Mann's Theorem and no new applications to classical problems had been obtained, though at the outset such applications were given by Šnirel' man.

This report suggests a new approach to the study of general additive regularities.

2. The special problem which we shall now discuss will give us an opportunity to introduce gradually the necessary new notions.

Let K be a finite set of k integers:

$$K = \{a_0, a_1, \ldots, a_{k-1}\}, \qquad a_i < a_{i+1}, i = 0, 1, \ldots, k - 2.$$

Let $T(M)$ be the number of elements of the finite set M and let $T = T(2K)$.

Obviously, $T \geqslant 2k - 1$. Actually, in the set $a_0 + K$ which consists of k numbers, the maximum is $a_0 + a_{k-1}$, whereas a set $a_{k-1} + K$ has this number as its minimum.

Let $T < Ck$, where C is a positive number. The problem is to determine the structure of K.

Here are some examples to clarify the question. If $T = 2k - 1$, then K is an arithmetical progression containing k numbers. In fact, if for some j with $0 \leqslant j \leqslant k - 3$ the inequality

$$a_{j+1} - a_j \neq a_{j+2} - a_{j+1}$$

holds, then the set $2K$ contains in addition to $2a_0, a_0 + a_1, 2a_1, a_1 + a_2, 2a_2, \ldots$ and $2a_{k-1}$ the number $a_j + a_{j+2}$ which is distinct from them.

It is possible to prove the following theorem by induction on k:

THEOREM 1.9. *If $0 \leqslant b < k - 2$ and $T = 2k - 1 + b$, then K is contained in the set*

$$K_a = \{a, a + q, a + 2q, \ldots, a + (k - 1 + b)q\},$$

where a is an integer and q a natural number.

Thus, if $T < 3k - 3$, then K is a subset of an arithmetical progression whose number of elements is not larger than $T - k + 1$.

Now will it make any difference if $T \geqslant 3k - 3$? More specifically: If we consider all arithmetical progressions containing the set K and choose from them an arithmetical progression with the minimum number of elements, then we can call this minimum the "length" of K. The question arises whether it is possible to evaluate the lengths of all K in the class of sets with given k and T as a function which depends only on the parameters.

The example

$$K = \{0, 1, 2, \ldots, k - 2, u\}, \qquad u > 2k - 4,$$

shows that this is impossible even for $T = 3k - 3$.

Nevertheless, results rather close to those obtained in Theorem 1.9 are true even in the case when $T \geqslant 3k - 3$. To formulate these results some new notions will be necessary, the most important one of them being a generalization of the notion of isomorphism between sets with algebraic operations.

3. DEFINITION. The subsets B' and C' of the sets B and C, respectively, with an algebraic operation, written additively, are called *isomorphic* if there exists a one-to-one mapping $B' \to C'$ which induces naturally a one-to-one mapping $2B' \to 2C'$. The mapping $B' \to C'$ is then called an *isomorphic mapping*, or an *isomorphism*.

An isomorphic mapping of a set K does not change the value of T. This is the reason why the description of the structure of K must be invariant under isomorphic mappings.

If $B' = B$, $C' = C$ and the mappings $B' \to C'$ and $2B' \to 2C'$ correspond to each other, our definition becomes the usual definition of an isomorphism between sets with an algebraic operation. In this case the addition of any number of elements corresponding to summands results in an element corresponding to the sum.

Let us consider the two additive semigroups $B' = \{0, 1, 2, \ldots\}$ and $C' = \{10, 11, 12, \ldots\}$. Then the mappings $B' \to C'$, defined by $u \to u + 10$, $u = 0, 1, 2, \ldots$, and $2B' \to 2C'$, defined by $u \to u + 20$, $u = 0, 1, 2, \ldots$, correspond to each other, and therefore B' and C' are isomorphic. This example shows that correspondences $B' \leftrightarrow C'$ and $2B' \leftrightarrow 2C'$ are sometimes incompatible.

If two given subsets are isomorphic, such a correspondence exists in general if the addition $K + K$ is not iterated. Thus the concept of isomorphism of subsets seems especially fitting for the study of questions of sums of finitely many sets, and this is the very thing necessary for the purposes of additive number theory in the study of problems with a bounded number of summands.

An analogy may be drawn between a local isomorphism between topological groups and an isomorphism between subsets ("algebraically local isomorphisms"). In the first case the isomorphism is restricted to a certain neighborhood, and in the second case by the number of operations.

Let B', $C' \subset E_n$ (the euclidean space) and assume that there exists a one-to-one correspondence of B' and C' such that any equation of the form $\bar{b}_1 - \bar{b}_2 = \bar{b}_3 - \bar{b}_4$ implies the corresponding equation $\bar{c}_1 - \bar{c}_2 = \bar{c}_3 - \bar{c}_4$, and $\bar{b}_1 - \bar{b}_2 \neq \bar{b}_3 - \bar{b}_4$ implies $\bar{c}_1 - \bar{c}_2 \neq \bar{c}_3 - \bar{c}_4$, where $\bar{b}_i \in B'$, $\bar{c}_i \in C'$ and $\bar{b}_i \leftrightarrow \bar{c}_i$, $1 \leqslant i \leqslant 4$. The conditions formulated above are necessary and sufficient for the sets B' and C' to be isomorphic.

4. FUNDAMENTAL THEOREM. *If $T < Ck$, $C \geqslant 2$, then there exist constants c and k_0, depending only on C, and a number $n \leqslant [C - 1]$ such that if $k > k_0$, then K is contained in a certain set K_0 of integers which is isomorphic to the set of interior points of some convex set $D \subset E_n$ and which satisfies the condition $T(K_0) < ck$.*

Here are some examples illustrating the Fundamental Theorem. If $2 \leqslant C < 3$, then $n = 1$, and D is an interval whose number of integral points is not larger than ck. Theorem 1.9 gives an elementary proof of a more powerful, similar result. If $3 \leqslant C < 4$ then $n \leqslant 2$. If $n = 1$, then K is contained in an arithmetical progression whose number of elements is not larger than ck.

If $n = 2$ then it is possible to show that D can be taken as a rectangle. Thus in this case K is contained in the union of some arithmetical progression with identical differences whose first elements form an arithmetical progression with another difference.

In fact, the lattice points of the rectangle which are situated on any one line parallel to the axis of abscissas are transformed into arithmetical progressions with identical differences. The image of the set of lattice points in the rectangle which are located on a certain line parallel to the axis of ordinates will be the set of the first elements of these progressions.

The Fundamental Theorem was proved with the help of a modification of the method of trigonometric sums.

5. Now it is possible to express a certain general viewpoint concerning the study of additive regularities. For this purpose we shall recall Klein's view on geometry as the science studying those properties of geometrical objects which are invariant under certain groups of transformations.

We are now able to state (see §3) that *the goal of additive number theory, while studying sums of sets, consists in the investigation of those properties of these sets which are invariant under isomorphic transformations.* From this point of view the study of relations between invariants of isomorphisms (henceforth called additive characteristics) appears to be very important.

The most important additive characteristics are T, the number R of distinct positive differences $a_i - a_j$ and the number $M = \int_0^1 |S|^4 \, d\alpha$, where $S = \sum_{j=0}^{k-1} e^{2\pi i \alpha a_j}$ (if $b_1, \ldots, b_T$ are the elements of the set $2K$ and r_s is the number of representations $a_i + a_j = b_s$, then $M = \sum_{s=1}^{T} r_s^2$).

Inverse problems of additive number theory (§1.7) may be stated as problems of describing sets with given invariants.

The Fundamental Theorem is the first step towards realizing the ideas expressed above. In formulating and proving it, one of the important tools used consists in isomorphic mappings of certain sets of integers into sets of lattice points of euclidean spaces of higher dimension. By using them it was possible to establish a fact which had naturally remained unnoticed previously, i.e. dimension is not invariant under isomorphic transformations.

It is worth mentioning that density is not invariant inasmuch as it may be changed by an isomorphic transformation, as is illustrated by the example

$$A = \{0, 1, 4, 8, 12, \ldots\}, \qquad \alpha'(A) = 1/4,$$

$$B = \{0, 1, 3, 6, 9, \ldots\}, \qquad \alpha'(B) = 1/3.$$

This explains why the application of the concept of density has certain limitations.

6. The analogy between Klein's view of geometry and the view of additive number theory as the theory of isomorphic transformations goes deeper then it may seem at first sight. It turns out that an isomorphic transformation is an affinity in a suitable euclidean space.

An isomorphic transformation of $K \subset E_m$ into E_n is called *nonsingular* if no hyperplane from E_n contains the image of K.

Suppose there exists a nonsingular transformation of the set K into E_n but no such transformation of K into E_{n+1}. Let this value of n be denoted by r. Then r is an additive characteristic of the set K.

THEOREM. *Let K_1 and K_2 be finite sets in E_r, neither of which is contained in any hyperplane from E_r. Then the sets K_1 and K_2 are isomorphic if there exists an affinity of E_r into itself which maps K_1 into K_2.*

7. *Applications.* Hinčin's Theorem is the special case of Mann's Theorem where two identical sets are added: If $\alpha'(A) = \alpha$ and $\alpha'(2A) = \gamma$, then $\gamma \geqslant \min(2\alpha, 1)$ provided that $a_0 = 0$.

An analog to Mann's Theorem for the addition of sets with positive asymptotic density was obtained by M. Kneser (1953). Expanding the theory of density, it became possible to show, roughly speaking, that if two identical sets are added then the density is at least doubled. It was found that, in general, the number 2 may be replaced by any positive constant C. This is the essence of the results derived from the Fundamental Theorem.

Let A be an increasing sequence of nonnegative integers:

$$A = \{a_0, a_1, \ldots, a_i, \ldots\}, \qquad a_{i+1} > a_i, i \geqslant 0,$$

$$d(A) = \lim_{k \to \infty} (a_1 - a_0, a_2 - a_0, \ldots, a_k - a_0).$$

Here $(a_1 - a_0, a_2 - a_0, \ldots, a_k - a_0)$ is the greatest common divisor of the numbers $a_1 - a_0, a_2 - a_0, \ldots, a_k - a_0$.

The number of positive integers in the set A (or $2A$) not exceeding x is denoted by $A(x)$ and $A_2(x)$, respectively. The asymptotic density $\alpha = \alpha(A)$ of the set A is defined by the equation

$$\alpha = \lim_{x \to \infty} \frac{A(x)}{x};$$

furthermore, we let $\gamma = \alpha(2A)$. The set of lattice points in the euclidean space E_n is denoted by Z_n. Let $\bar{e}_1, \ldots, \bar{e}_n$ be an orthonormal basis of the space E_n. Let φ: $Z_n \to Z_1$ be a homomorphism satisfying $\bar{e}_i \varphi = 0, 1 \leqslant i \leqslant n - 1$, and $e_n \varphi = 1$. Finally, let $N^{(x)}$ be the subset of the given set $N \subset E_n$ consisting of those elements for which $x_n \leqslant x$. As a consequence of the Fundamental Theorem the following theorem is proved in Chapter III:

THEOREM. *Suppose the set A satisfies the following conditions: $a_0 = 0$, $d(A) = 1$, $\alpha > 0$, $\gamma > C\alpha$ and $C > 3/2$. Furthermore, assume that there exists a constant C_1 such that the inequality $a_{i-1} > C_1 a_i$ is satisfied only for a finite number of indices i. Then there exist a cylinder $N \subset E_n$, $n \geqslant 2$, whose base is a convex set $P \subset L(\bar{e}_1, \ldots, \bar{e}_{n-1})$, and positive constants c and α_0, depending on C and C_1 only, such that for $\alpha < \alpha_0$ the following assertions are true:*

(1) $A \subset (N \cap Z_n)\varphi$.
(2) $N \cap Z_n$ and $(N \cap Z_n)\varphi$ are isomorphic.
(3) $V(P) < c\alpha$, $T(N^{(x)} \cap Z_n) < c\alpha x$.
(4) $n \leqslant [2C - 1]$.

For sequences of zero density an analog of Hinčin's Theorem was conjectured by P. Erdös [13]. His conjecture states that

$$\delta = \varliminf_{x \to \infty} \frac{A_2(x)}{A(x)} \geqslant 3, \quad \text{if} \quad \lim_{x \to \infty} \frac{A(x)}{x} = 0.$$

By means of the Fundamental Theorem the following more general assertion was proved:

THEOREM. If $\varliminf_{x \to \infty} A(x)/x = 0$ and

$$\varlimsup_{x \to \infty} \frac{A(x)}{x} < \frac{1}{2d(A)}$$

then $\delta \geqslant 3$.

The Fundamental Theorem also leads to applications to sum-sets of sets of residues modulo a prime, thus strengthening results received by Cauchy, Davenport and Vosper, and for point sets of positive measure (strengthening the Brunn-Minkowski Inequality).

8. In the Fundamental Theorem the structure of a set K with small double set (i.e. with $T < Ck$) is investigated. Further possible lines of investigation are the following ones:

Study the structure of K when $T/k \to \infty$.

Generalize the problem to the case where several summands are considered (iterated addition) or when the summands are different.

Investigate the structure of K when different additive characteristics of the set K or their combinations are given.

After dividing the sets consisting of a finite number of lattice points into classes of isomorphic sets, study the order of increase of the function $t(k)$, i.e. the number of such classes for given k.

All these problems may be studied for sets of a more general nature than sets of lattice points, especially in groups, both abelian and nonabelian, which would be of special interest.

In this index reference is made to the following books:

[C] J.W.S. Cassels, *An introduction to the geometry of numbers*, Die Grundlehren der Math. Wissenschaften, Band 99, Springer-Verlag, Berlin, 1959; Russian transl., "Mir", Moscow, 1965. MR **28** # 1175; **31** # 5841.

[K] A.G. Kuroš, *Lectures on general algebra*, Fizmatgiz, Moscow, 1962; English transl., Chelsea, New York, 1963; Internat. Series of Monographs in Pure and Appl. Math. vol. 70, Pergamon Press, Oxford, 1965. MR **25** # 5097; **28** # 1228; **31** # 3483.

[N] I.P. Natanson, *Theory of functions of a real variable*, GITTL, Moscow, 1950; 2nd rev. ed., 1957; English transl., vols. I, II, Ungar, New York, 1955, 1961. MR **12**, 598; **16**, 120, 804; **26** # 6309.

All page references are to the English editions; for Kuroš the first page number refers to the Chelsea translation; the second, to the Pergamon.

Algebraic operation K30, 20

Basis of a lattice C9

Class of isomorphic sets 5

Density of a sequence
 asymptotic 75
 Šnirel'man 75

Dimension of a point set 24

Equivalence relation K18, 7

Farey dissections 52

Farey sequence 52

Hyperplane 24
 supporting 59

Integral vector 23

Interval of a fraction 52

Invariant
 of an isomorphic mapping 40
 of order s 45

Ismorphism
 between subsets of a set with an
  algebraic operation 2, 3
 of order s between subsets 4
 of two finite classes of sets 4

Kernel of a homomorphism K103,102

Lattice C9

Lebesgue (outer) measure N I 63, 67

Length of a set 14

Mapping
 isomorphic 2
 monomorphic 43
 naturally induced 2
 regular isomorphic 29

Monomorphic mapping of subsets 43

Parallelepiped
 canonical 53
 fundamental, of a lattice C196

Plane 24

Projection onto a plane 27

Semigroup K30, 20

Set
 closed N II 61
 convex C2
 open N II 62

Straight line 24

Sum of subsets (sets) 1

Torsion-free group K44, 36

Volume
 of a set of lattice points 30
 reduced 35

BIBLIOGRAPHY

1. S. N. Bernšteĭn, *Theory of probability*, OGIZ, Moscow, 1946. (Russian)
2. A. Beurling, *Analyses de la loi asymptotique de la distribution des nombres premiers généralisés*. I, Acta Math. **68** (1937), 255–291.
3. B. M. Bredihin, *Free numerical semigroups with power densities*, Dokl. Akad. Nauk SSSR **118** (1958), 855–857. (Russian) MR **20** #5175.
4. ———, *Free numerical semigroups with power densities*, Mat. Sb. **46 (88)** (1958), 143–153. (Russian) MR **21** #87.
5. ———, *Elementary solutions of inverse problems on bases of free semigroups*, Mat. Sb. **50 (92)** (1960), 221–232. (Russian) MR **26** #246.
6. ———, *The remainder term in the asymptotic formula for $\nu G(x)$*, Izv. Vysš. Učebn. Zaved. Matematika **1960**, no. 6 (19), 40–49. (Russian) MR **26** #92.
7. J. W. S. Cassels, *An introduction to Diophantine approximation*, Cambridge Tracts in Math. and Math. Phys., no. 45, Cambridge Univ. Press, New York, 1957. MR **19**, 396.
8. ———, *An introduction to the geometry of numbers*, Springer-Verlag, Berlin, 1959; Russian transl., "Mir", Moscow, 1965. MR **28** #1175; **31** #5841.
9. A. Cauchy, *Recherches sur les nombres*, J. École Polytechn. **9** (1813), 99–123.
10. H. Davenport, *On the addition of residue classes.*, J. London Math. Soc. **10** (1935), 30–32; Russian transl., Uspehi Mat. Nauk **7** (1940), 90–92.
11. P. Erdös, *On the asymptotic density of the sum of two sequences one of which forms a basis for the integers*. II, Trudy Tbilissk. Mat. Inst. **3** (1933), 217–223.
12. ———, *On an elementary proof of some asymptotic formulas in the theory of partitions*, Ann. of Math. (2) **43** (1942), 437–450. MR **4**, 36.
13. ———, *Some unsolved problems*, Magyar Tud. Akad. Mat. Kutató Közl. **6** (1961), 221–254; Russian transl., Matematika **7** (1963), no. 4, 109–143. MR **31** #2106.
14. G. A. Freĭman, *Inverse problems of additive number theory*, Učen. Zap. Kazan. Univ. **115** (1955), no. 14, 109–115. (Russian) MR **18**, 112.
15. ———, *Inverse problems of the additive theory of numbers*, Izv. Akad. Nauk SSSR Ser. Mat. **19** (1955), 175–284. (Russian) MR **17**, 239.
16. ———, *The addition of finite sets*. I, Izv. Vysš. Učebn. Zaved. Matematika **1959**, no. 6 (13), 202–213. (Russian) MR **23** #A3684.
17. ———, *Inverse problems of additive number theory*. IV. *On the addition of finite sets*. II, Elabuž. Gos. Ped. Inst. Učen. Zap. **8** (1960), 72–116. (Russian)
18. ———, *Inverse problems of the additive theory of numbers. On the addition of sets of residues with respect to a prime modulus*, Dokl. Akad. Nauk SSSR **141** (1961), 571–573 = Soviet Math. Dokl. **2** (1961), 1520–1522. MR **27** #5744.
19. ———, *Inverse problems of additive number theory*. VI. *On the addition of finite sets*. III, Izv. Vysš. Učebn. Zaved. Matematika **1962**, no. 3 (28), 151–157. (Russian) MR **27** #2464.
20. ———, *Inverse problems of additive number theory*. VII. *The addition of finite sets*. IV. *The method of trigonometric sums*, Izv. Vysš. Učebn. Zaved. Matematika **1962**, no. 6 (31), 131–144. (Russian) MR **26** #2420.

21. ———, *Inverse problems of additive number theory*, Proc. Fourth All-Union Math. Congr. (Leningrad, 1961), vol. II, "Nauka", Moscow, 1964, pp. 142–146. (Russian) MR **36** #3743.

22. ———, *Inverse problems in additive number theory. VIII. On a conjecture of P. Erdös*, Izv. Vysš. Učebn. Zaved. Matematika **1964**, no. 3 (40), 156–169. (Russian) MR **29** #5790.

23. ———, *Inverse problems in additive number theory. IX. The addition of finite sets. V*, Izv. Vysš. Učebn. Zaved. Matematika **1964**, no. 6 (43), 168–178. (Russian) MR **31** #148.

24. ———, *Addition of finite sets*, Dokl. Akad. Nauk SSSR **158** (1964), 1038–1041 = Soviet Math. Dokl. **5** (1964), 1366–1370. MR **29** #5791.

25. R. Henstock and A. M. Macbeath, *On the measure of sum-sets. I. The theorems of Brunn, Minkowski and Lusternik*, Proc. London Math. Soc. (3) **3** (1953), 182–194. MR **15**, 109.

26. A. Ja. Hinčin, *Sur la sommation des suites d'entiers positifs*, Mat. Sb. **6** (48) (1939), 161–166. (Russian) MR **1**, 201.

27. ———, *Three pearls of number theory*, OGIZ, Moscow, 1947; English transl., Graylock Press, Albany, N.Y., 1952. MR **11**, 83; **13**, 724.

28. I. H. B. Kemperman, *On small sumsets in an abelian groups*, Acta Math. **103** (1960), 63–88. MR **22** #1615.

29. M. Kneser, *Abschätzung der asymptotischen Dichte von Summenmengen*, Math. Z. **58** (1953), 459–484. MR **15**, 104.

30. ———, *Ein Satz über abelsche Gruppen mit Anwendungen auf die Geometrie der Zahlen*, Math. Z. **61** (1955), 429–434. MR **16**, 898.

31. A. G. Kuroš, *Lectures on general algebra*, Fizmatgiz, Moscow, 1962; English transl., Chelsea, New York, 1963; Internat. Series of Monographs in Pure and Appl. Math., vol. 70, Pergamon Press, Oxford, 1965, MR **25** #5097; **28** #1228; **31** #3483.

32. H. B. Mann, *A proof of the fundamental theorem on the density of sums of sets of positive integers*, Ann. of Math. (2) **43** (1942), 523–527. MR **4**, 35.

33. ———, *Addition theorems: The addition theorems of group theory and number theory*, Interscience, New York, 1965. MR **31** #5854.

34. I. P. Natanson, *Theory of functions of a real variable*, GITTL, Moscow, 1950; English transl., Ungar, New York, 1955. MR **12**, 598; **16**, 804.

35. H. Ostmann, *Additive Zahlentheorie. I. Allgemeine Untersuchungen*, Ergebnisse der Mathematik und ihrer Grenzgebiete, Heft 7, Springer-Verlag, Berlin, 1956. MR **20** #5176.

36. ———, *Additive Zahlentheorie. II. Spezielle Zahlenmengen*, Ergebnisse der Mathematik und ihrer Grenzgebiete, Heft 11, Springer-Verlag, Berlin, 1956. MR **20** #5176.

37. P. P. Postnikova, *Fluctuations in the distribution of fractional parts*, Dokl. Akad. Nauk SSSR **161** (1965), 1282–1284 = Soviet Math. Dokl. **6** (1965), 597–599. MR **31** #5857.

38. K. F. Roth, *On certain sets of integers*, J. London Math. Soc. **28** (1953), 104–109. MR **14**, 536; 1278.

39. E. Schmidt, *Die Brunn-Minkowskische Ungleichung und ihr Spiegelbild sowie die isoperimetrische Eigenschaft der Kugel in der euklidischen und nichteuklidischen Geometrie. I*, Math. Nachr. **1** (1948), 81–157. MR **10**, 471.

40. B. I. Segal, *Trigonometric sums and some of their applications in the theory of numbers*, Uspehi Mat. Nauk **1** (1946), no. 3–4 (13–14), 147–193. (Russian) MR **10**, 17.

41. L. G. Šnirel' man, *On additive properties of numbers*, Izv. Politehn. Inst. **14** (1930), no. 2/3, 3–28. (Russian)

42. V. H. Tašbaev, *An inverse additive problem*, Mat. Sb. **52** (94) (1960), 947–952. (Russian) MR **24** #A1900.

43. A. G. Vosper, *The critical pairs of subsets of a group of prime order*, J. London Math. Soc. **31** (1956), 200–205. MR **17**, 1056.